Why Genes Are Not Selfish and People Are Nice

'This wise and far-reaching book points the way to a better, more inclusive kind of science and a better, more inclusive kind of religion in a positive, constructive relationship. This is surely what we need most in the twenty-first century and Tudge is a genial guide for all who feel the need to move on from scientific and religious fundamentalism, environmentally destructive capitalism and an economic philosophy of selfishness, competition and limitless growth. Tudge points the way to a new kind of agriculture, a new way of living in harmony with our planet and the universe, and with each other. This book is an impressive synthesis and is admirably non-technical, conversational and approachable. Tudge, one of our most distinguished science writers, is a prophet for our time, and a very welcome voice of sanity and reason.'

RUPERT SHELDRAKE

~

'Colin Tudge is a scientist with a difference; a scientist who is not afraid of embracing transcendence. He sees no conflict between science and wisdom. This profound and revolutionary book is a deep exploration of science, life and transcendence. It is a courageous critique of materialism and a persuasive argument in favour of fresh thinking. Tudge challenges us to come out of our narrow view of science and look at life with a much bigger perspective.'

SATISH KUMAR

~

'This book more than lives up to its subtitle. It does indeed challenge big bad ideas, whether they be about the natural world, the human condition within it, or our habits of thought and behaviour, and suggests some bigger, better ideas for the future. In short replace the conventional wisdom. All this is laid out in easy but scholarly fashion, and the conclusions are a personal testament. Think differently is the message. We are now better able to do so.'

SIR CRISPIN TICKELL

WHY GENES ARE NOT SELFISH AND PEOPLE ARE NICE

A Challenge to the Dangerous Ideas that Dominate Our Lives

COLIN TUDGE

Floris Books

This book is also available as an eBook

British Library CIP Data available
ISBN 978-086315-963-3
Printed in Great Britain
by TJ International Ltd, Cornwall

Contents

Preface

I have been writing this book in my head for more than sixty years. That is, I realised at the age of six that I wanted to be a biologist (although I don't think I used words like 'biologist' at that time) for I was, as I think most budding biologists are, besotted by nature; by the feeling that life is endlessly absorbing but also that it is magical. It has the quality that the Lutheran theologian Rudolf Otto called 'numinous' – although I didn't learn that word either, until very much later. I was lucky enough to be born at a time when excellent education was free, and went on to a school where science was excellently taught, and from there to 'read' zoology at a very old and cold university that had somehow managed to keep its spires above the encircling swamp, and there too the teaching was outstanding. But that was in the 1960s and we tended to pursue the ludicrous idea that all life could be reduced to physics, and physics to maths, and that was the end of it.

Yet I never quite felt that that was the end of it. I always had an ill-formed but nonetheless powerful feeling that there is a great deal more to life and the universe than meets the eye – or ever can meet the eye, no matter how much science we may do. I never quite managed to become a Christian but I always felt (despite my youthful denials) that religion was saying important things, although I wasn't quite sure what. I always felt too – in fact I think I took it to be self-evident – that St Francis was right: that other creatures are truly our fellows; and this feeling was constantly reinforced by my lifelong absorption in the idea of evolution. I also felt that the Founding Fathers of the US were stating the obvious in the second paragraph of their Declaration of Independence when they said that 'all men are created equal'. Of course they are. How could it be otherwise?

Then at about age seventeen I started reading Aldous Huxley, and in particular his magnificent saga *Point Counterpoint*; and so had my first intimation that all knowledge is all one – or at least that it could and should be one. The fragmentation into 'science', 'philosophy', and 'religion' was just for convenience, and it was a sad and dangerous thing to leave them in their separate packages (or indeed on separate campuses). Later I learned that everything could and should be brought together under the grand heading of metaphysics; and that the unified vision that metaphysics can provide was needed to underpin the practicalities of politics and economics, and hence of our day-to-day lives. Metaphysics brings coherence (and nothing else does).

All these ideas kicked around in my head for about half a century until finally, about three years ago, I decided it was time to pull things together before the grim reaper came a-knocking. So here's the result. It isn't a finished, definitive work because in an endeavour like this there can be no such thing. It is, however (I hope) an agenda for discussion. The very concept of 'metaphysics' has gone missing, and as the Islamic scholar Seyyed Hossein Nasr has commented, the loss of metaphysics from the western world might be seen as the root cause of all our ills. That loss, after all, means the end of unified thinking. My ambition, then (besides the wish to consolidate the ponderings of a lifetime) is to put metaphysics back where it belongs, at the heart of all human thinking and all human affairs.

I am aware of my debt to an enormous number of people – far too many to list, for much of what all of us learn is by osmosis, entirely informally, often through chance remarks. I must though mention my teachers at school and university, for teachers really do set the tone. Then in my early journalistic days I had particularly illuminating conversations with Peter Bunyard, Bernard Dixon, Geoff Watts, and Fred Kavalier. I am very grateful too to Helena Cronin, who in the 1990s invited me to join the Darwin Centre at the London School of Economics as a Visiting Research Fellow, where I met some of the world's finest biological thinkers. I also had good conversations with Sophie Botros who introduced me to some key notions of philosophy. More recently, since we moved to Oxford, my wife Ruth and I have spent many a pleasant and instructive evening with Paul and Chris Wordsworth, and John and Sally Lennox, who combine medicine and maths with devout Christianity. I have also enjoyed good discussions with my old friend Georgina Ferry, on science in Elizabethan times. Over the years, too, I have learned a great deal from Rupert Sheldrake and Robert Temple; and from Tim Bartell, who put me right (I hope) on the structure of moral philosophy. At Oxford I have benefited enormously from the Ian Ramsey lectures on science and religion, and especially from (albeit brief) conversations with John Hedley Brooke. Through Aaron and Sid Cass I have got to know many of the people at the Beshara School at Chisholme on the Scottish borders, where metaphysics is discussed with particular emphasis on the teaching of Ibn 'Arabi. I have also been lucky enough over the past few years to form an association with Schumacher College at Dartington in Devon, and in particular with Satish Kumar.

Finally I am very grateful to Sir Crispin Tickell, Rupert Sheldrake, and my nephew John Harris who read an earlier draft of this book and gave me valuable advice; and to Christopher Moore at Floris Books who has done much to help me tidy it up. Most of all however I am grateful to my wife, Ruth (West), who looks after me remarkably well and is also in some relevant areas a far better scholar than I am, and a very astute critic.

Colin Tudge, November 2012

Introduction: The Ideas In The Basement

I want to present a different view of life – compounded of ideas that in most important respects are completely opposed to the kinds of beliefs that now dominate the modern western world and so have come to dominate the world as a whole. For we in the western world have misconstrued – well, just about everything: the nature of the universe, of life, and of our own human selves; the nature of truth; and the nature of right and wrong. From these fundamentals the errors compound: we treat the world badly, we treat each other badly, we put tremendous store by ideas that are decidedly flimsy and afford the people who promote those ideas far too much respect. To be more specific, our politics is unjust, our economic system borders on the insane, our governments for the most part are not on our side, our science which should be our great liberator has become the handmaiden of big business, while religion is all over the place, fractious and narrow-minded where it should be all-embracing.

The ideas that dominate and underpin our lives are the stuff of the *Zeitgeist*, the spirit of the age. They lie in the basement of our minds, taken for granted, rarely brought out for inspection. If we did drag them out into the light, we would find that very few of them are truly 'robust'. Few have any solid evidence behind them. Most have simply been inherited, like folklore. Many, we would find, are just plain wrong. When they are laid side by side, too, we can see that they do not cohere. The ideas that form the *Zeitgeist,* the core of our beliefs, are a jumble, as the contents of the basement commonly are. Yet we live our lives by them. No wonder people feel uneasy. No wonder the world is in a mess.

Part of the point of this book is to rummage through the basement: to explore the ideas of the modern *Zeitgeist*, and see if they really stand up. Most of them, I intend to show, do not. Many are just plain wrong, or at least need serious rethinking. But also, much more importantly, I want to present the alternative views – which tend to be in complete opposition to what is now commonly believed, and yet are at least as likely to be true. I also want to show, as a not inconsiderable bonus, that the ideas that are actually true are not a jumble. Together they form a worldview that is perfectly coherent. Furthermore – and this really is serendipitous – a worldview rooted in the ideas that are most likely to be true could lead us to create a world that is truly convivial, which our descendants and our fellow creatures could enjoy for the next million years. Then they could draw breath and contemplate the following million.

So this book emerges as an exercise in metaphysics; for the goal of metaphysics is to find out what is true, and what is good, applying all our faculties to the task – both our intellect and our intuitions; and to bring all truths together into a single worldview that is both all-embracing and coherent. As the Islamic scholar Seyyed Hossein Nasr points out in *Man and Nature,* the loss of metaphysics from western thinking – even the word itself has largely disappeared – is at the root of all our troubles, not just in the west but in the world as a whole, because the west is so dominant. The rot set in (I suggest) with the European Renaissance, from the fifteenth century onwards. Before that, in the Middle Ages, all thought was intertwined: theology, philosophy, and early forms of science. But the Renaissance led to fragmentation – reflected in the design of modern universities, where science, theology, philosophy, and all the 'humanities' are pursued in discrete 'faculties', which sometimes occupy separate campuses. Each produces its own species of intellectual who for the most part spend the rest of their lives in mutual misunderstanding, occasionally brought together into 'think-tanks' that are supposed to provide overviews of life's many problems, but seem doomed by their disparateness to do no such thing. Of course, in order to raise science and other modes of thought to the greatest heights, each of them needs to some extent to be pursued in isolation, without noises off. But none should ever be allowed to lose sight of the others. All of them, all the time, should be seen to be embedded in an all-embracing metaphysic.

So what are these big ideas that underpin our lives, and need to be re-thought? The main ones, I suggest, are as follows:

The big ideas

Most obviously we must ask – 'What is the universe really like?' This you may feel is just a matter of science – which up to a point is true, but with several caveats. For one thing, science has a horrible way of binding itself to dogmas from which it cannot easily escape. Notably, this past 150 years, a group of biologists known as 'neo-Darwinists' – or even, sometimes, 'ultra-Darwinians' – have picked up on Charles Darwin's idea that natural selection is driven by competition and concluded that all life is inveterately competitive; that nature is as Lord Tennyson said, 'red in tooth and claw'. This idea now dominates the western world. It is the lynch stone of the global, neoliberal, all-against-all economy. But actually, it's not true.

More broadly, science does what science does very well – it analyses the material world. But it cannot ask the two most fundamental questions of all. It cannot sensibly ask, 'Is the material world all there is, or is there more to life and the universe than meets the eye?' Neither can science tell us, 'How come?' Thus a scientist may explain in minutest detail what happened in the microseconds after the Big Bang – but how did it all come about? How and why was it that things were as they were, such that the Big Bang happened? Some scientists in the manner of logical positivists simply declare that such questions are silly *because* they cannot be answered by science. But most people feel deep down that this is a duck-out – and, I suggest, most people are right.

Yet scientists of the hard-nosed kind are wont to insist that what they cannot study is simply not real, and since they are equipped only to study the material world they conclude that there can be nothing else. Materialism prevails, therefore; meaning, in the words of the Roman poet Lucretius that the universe consists only of 'atoms and the void'. With materialism goes the idea that there is no purpose in the universe, and no direction. As the Oxford neo-Darwinist materialist Richard Dawkins has written:

> The universe we observe has precisely the properties we should expect if there is, at bottom, no design, no purpose, no evil and no good, nothing but blind pitiless indifference.[1]

To insist that the universe is just material 'stuff', and nothing else, is to reject outright the concept of transcendence – for transcendence, defined crudely, says that there is more to the universe than meets the eye, or ever can meet the eye, no matter how much science we may do. Transcendence is the notion that underpins and unites all religions – for even those like Buddhism and Confucianism, which don't recognise the specific concept of 'God', do insist that the material world is not all there is, and indeed is just the surface of things. So out-and-out materialists must reject a key notion of religion and so perforce are atheists – of which Professor Dawkins has become the world's most famous exponent (even though in a discussion with Archbishop Rowan Williams in February 2012 he declared himself to be 'agnostic'). Dawkins and his fellow materialist atheists claim to be ultra-rationalist and therefore to be party to the truth. But they have mistaken the nature of rationalism; and although rational thinking is of course vital it is not the royal road to truth, and in the end, the materialist-atheism that we are told is so hard-headed is nothing more than dogma (which is ironical, given that the atheists express contempt for the dogmas of religion).

Then we are told that since the material world is all there is, and since scientists are so good at studying it, that we can reasonably put all our faith in science and in the 'high' (science-based) technologies that come out of it. After all, as the Oxford chemist Peter Atkins has told us:

> Science, the system of belief founded securely on publicly shared reproducible knowledge, emerged from religion. As science discarded its chrysalis to become its present butterfly, it took over the heath. There is no reason to suppose that science cannot deal with every aspect of existence. Only the religious – among whom I include not only the prejudiced but also the underinformed – hope there is a dark corner to the universe, or of the universe of experience, that science can never hope to illuminate. But science has never encountered a barrier, and the only grounds for supposing that reductionism will fail are pessimism on the part of scientists and fear in the minds of the religious.[2]

But this is an illusion. For one thing, it's clear, now, that the universe is innately unpredictable. Simple causes do not lead to simple effects, as Isaac Newton and his Enlightenment successors imagined. Now we can see that in the real universe as opposed to the equations of mathematics, cause and effect are 'non-linear'; which means in practice that although some outcomes are more likely than others, no-one can ever know what will really happen. Life, which might be seen as the ultimate physical expression of the universe, is non-linear in spades. In more philosophical vein, it's clear now that the explanations that science provides, brilliant and useful though they may be, are not cast-iron, and never can be, and even if they were we could never know that this is so. More broadly, the main reason that scientists seem so successful is that they are careful to address only those questions that they think they can answer. Faith in science as the path to omniscience is seriously misguided.

Yet faith in science has led to the even more pernicious faith in high technology – the belief that it will make us omnipotent; that one day we will control or 'conquer' all of nature, and are well on the way do doing so. This idea is used to justify the prevailing economic strategy, based on endless material growth. For in practice the unswerving pursuit of growth does a huge amount of damage, not least to the climate, and this could wreck us all. But the governments and corporates and banks who run the global economy take it to be the case that whatever damage we may do along the way, however deep the holes in which we may dig ourselves, we can always dig ourselves out again with high tech. Alas there are scientists on hand to tell them that this is so. But in the end, nature is beyond our ken and well beyond our control, except temporarily, in little bits, here and there; just enough to foster our illusions.

Then we have ideas about who we are and where we stand in relation to other creatures and the universe as a whole; and prominent among those ideas in the western world (though not in many another society) has been the mortal fear of anthropomorphism, and the absolute acceptance of anthropocentricity. Anthropomorphism means to ascribe human characteristics to non-human creatures, or even to inanimate objects, and this has generally been taken to be a thoroughly bad thing to do. Anthropocentricity is the idea that human beings are the only creatures that really count, and the rest may be treated as resources – and this, nowadays, is taken to be self-evident. Both ideas are clearly very ancient. Indeed they seem to be implied by Genesis 1:26:

> And God said, Let us make man in our image, after our likeness: and let them have dominion over the fish of the sea, and over the fowl of the air, and over the cattle, and over all the earth, and over every creeping thing that creepeth upon the earth. (King James Bible.)

In the three Abrahamic religions, Judaism, Christianity, and Islam, the notion that we are made in God's image has often been taken to imply that other creatures are not; while the word 'dominion' has commonly been taken to mean that we have the right to shove the rest around. Many other religions, and many thinkers within the Abrahamic religions, take a quite different view. Some see all creatures, and indeed everything in the universe, as manifestations of God – so everything in a sense is 'in his image'. Many theologians have pointed out too that 'dominion' is not a good translation. It should be 'stewardship' (although some feel that that too is presumptuous. I feel that stewardship is necessary, but the task must be approached with extreme humility).

Be that as it may, modern biologists and particularly of the kind known as behaviourists have regarded anthropomorphism as the great no-no, and gone to enormous lengths to widen the conceptual gap between us and them; and many modern scientists, economists, and politicians in anthropocentric vein are content to treat other creatures as resources. Again these ideas are deeply pernicious and when we explore them we find they are unfounded. Pleasingly, however, in truly modern biology as opposed to the hard-nosed kind of the mid-twentieth century, anthropomorphism has gained a respectable place.

We have also allowed ourselves, these past few thousand years, to be persuaded that we, human beings, are a bad lot, not to be trusted. So it is that Christianity, which in many ways is so attractive, has burdened itself and the rest of us with the concept of 'original sin' for which we must spend our lives atoning – although what exactly is the nature of the sin and why it is so dire is not at all clear, at least to me (or to many a Christian of my acquaintance). Some modern neo-Darwinians have suggested in similar vein that human beings must be bad since as living creatures we are bound to compete, head to head from conception to the grave, which implies that we are condemned by our own evolution to do each other down.

The notion that we are all vicious at our core has in turn allowed priests and politicians to lord it over us. So it was that

Britain's Lord (Roy) Hattersley, although he is seen as one of the last of the Labour Party socialists, said recently on BBC radio that without 'strong government' to keep us in check the rest of us (in the old days known as 'the mob') would tear each other to pieces (though I paraphrase). By such conceits all hierarchies including or especially the most oppressive insist that they are necessary; that for our own good we need special people to tell us what to do (and that they, the people in power, are indeed the appropriate people). We allow them to get away with this.

Overall, as the *coup de grâce*, we have a ludicrous concept of progress – or at least have allowed ourselves to be dominated by people whose vision of the future, if vision it can be called, is crass. Thus progress is conceived in materialist terms. The number one goal of all present governments is to generate wealth, irrespective it seems of how that wealth is created, or who hangs on to it, or what it is used for. The rising tide of wealth is called 'economic growth'. Progress, too, is perceived as an exercise in control. Nature as a whole must be treated as a resource which in turn can be turned into commodities to be sold for money. All human affairs, the minutiae of our lives, must be documented and cross-referenced – so progress emerges a giant exercise in bureaucracy, reinforced by high-tech surveillance. People who won't subscribe to this view of progress are written off as hippies or backsliders or hopeless romantics, and countries that fail to follow the path that has been chosen for them are deemed to have 'failed'. If they have no oil or fertile land they are simply ignored. If they have any of either (or tin or tungsten or whatever it may be) a pretext will be found to take them over (for their own good).

In short, to put the world to rights, and for our own peace of mind, we need to drag all our accepted and unexamined ideas out into the light for re-examination; and use all our faculties, both our powers of reasoning and our intuition, to see what is really true or at least, what is worth taking seriously. This is metaphysics. If we do the job properly, this will lead us to a new and altogether more rounded and satisfactory view of life; and we can build on that new worldview to create a far, far better world than we have now.

If we go on as we are, burdened by ideas that are deeply suspect and to a large extent vile, then we and our fellow creatures have very little chance of surviving even through the next century in a tolerable form. But if we get our ideas straight, and in line with our true feelings, and act accordingly, then our descendants should still be here in a million years time, and still enjoy the company of our fellow creatures. The stakes are very high.

PART I

THE NATURE OF NATURE

1. What Darwin Really Said And Why He Said It

As 2012 lurches towards its close, with the global economy still mired in depression, while civil war is threatening or already in full spate in the Middle East, and a billion people (one in seven) are permanently hungry, and another billion are sick from eating too much, and a billion more (nearly a third of all the world's city-dwellers) live in urban slums, and our fellow species pass into oblivion more rapidly than ever before, and the climate is changing radically in ways we cannot in detail predict, the overriding theme of global politics is – *competition!* The task is not to grow food, or to help each other out in what perhaps is the most critical time in the history of humanity, or to take care of our beleaguered fellow creatures, but for each country and corporate and individual to pursue their own narrow interests and get ahead of the rest. Somehow, if they do (fingers crossed!) everything will turn our alright; and if it doesn't, well, that's life.

To the naïve observer this may well seem ludicrous. But there is worse. For this ultra-crude political philosophy, and the ultra-crude economic theory of neoliberalism in which it is made manifest, is commonly perceived to be supported by science; and science, we have been brought up to believe, can't be wrong, because science is 'rigorous', and scientists are not swayed by anything so flakey as emotion, prejudice, or vested interest. The particular bit of science that's brought to bear is known as 'neo-Darwinism' which, allegedly and to some extent in fact, derives from the seminal work of Charles Darwin of 1859 – *On the Origin of Species by Means of Natural Selection.* Darwin (so the popular memory has it) *proved* that competition is life's great driver: the central fact of nature, and indeed of life. Moral philosophers (at least from the time of St Paul) have been pointing out that what is natural is not necessarily good, but this caveat is more or less ignored. The *Zeitgeist* has it that competition is good and necessary and in any case is inevitable, because it is natural. So we have a global economy rooted in the notion of all-against-all, and politics to match (even though, of course, countries still find it necessary to form alliances, and companies form cartels, the better to biff the rest; and the most powerful governments are now intertwined with the most powerful corporates to form what might be called a 'government-corporate complex'); and all is rooted in the supposition that this is the way life is, so we can't do things any other way.

But of course the naïve observer is right. Naïve observers often are, because they rely on intuition, and we should trust our intuitions. Competition is a fact of life but it is not the central fact. It is not the driver. The world would not grind to a halt without it. Absolutely not should the perceived need to compete be the basis of our economy, or of global politics. That is bound to be destructive, and it is. Neoliberal supporters of the ruthless global economy do indeed claim that their ideas are rooted in science but the science they are rooted in is seriously old-fashioned. Darwin was indeed a genius (and a lifelong hero of mine, for what the information is worth). Together with his near contemporary, Gregor Mendel, who gave the first convincing account of heredity, Darwin can properly be said to have begun the modern age of biology; and the ideas of the two together have spilled well beyond biology, and into all corners of western thought.

But Darwin, like all of us, inherited several thousand years of intellectual and spiritual baggage. Like all of us, too, he was a child of his time – and his time, the early nineteenth century, was particularly threatening; remarkably similar to our own, in fact, although without the benefit of quite so much hindsight (not that we seem to make much use of it). This inheritance, and his personal experience, coloured his view of life. For although he was himself a gentle man, a liberal, a humanitarian, and one of the greatest field naturalists of all time with a deep, true love of other creatures (including, as J.B.S. Haldane said of God, 'an inordinate fondness for beetles') the picture he presents to us of the natural world is gloomy, not to say rather terrifying.

Several thousand years of pessimism

Western culture is traditionally said to be rooted in that of the Greeks on the one hand, and the Jews on the other; and although that's a huge simplification, it is surely true up to a point. Unfortunately, both of those great cultures tended to take a pretty dim view of life in general, and of humanity in particular. The lives of Homer's heroes and of their society were subject to the whims of an entire menagerie of gods – a fickle and self-centred lot, bickering and sulking and prone to infanticide and patricide; qualities reflected in the all-too-human, petulant and vengeful heroes of Greek literature. Plato, a few centuries after Homer, had more or less shaken off the old gods but he still took a pretty dour view of earthly life and of humanity at large. He took it to be self-evident that the majority of us, obviously including women and slaves, were not to be trusted with anything that required serious thought (and poets should be banned lest they fire the imagination of those whose imaginations are best left unfired). People at large had to be kept in their place by the patrician intelligentsia – people like himself – who were duty-bound to take charge.

The Jews worshipped a single God sometimes known as Jehovah (after a few flirtations with various graven images) but Jehovah, at least as represented in the early books of the Old Testament, sometimes seems on first reading to be just as hard to deal with as the gods of the Greeks. So he made human beings in his own image, as intelligent creatures with free will – but then booted them out of Paradise as soon as they began to indulge their God-given curiosity, and ate from the tree of the knowledge of good and evil. Ever since the booting-out (as St Augustine explained in the fifth century) human beings (traditionally known as 'Man') have been obliged to atone for this 'original sin' (although it is never *entirely* clear what the sin was). In the sixteenth century, John Calvin and Martin Luther, good Catholics both who in their attempts to reform Catholic practice managed to give rise to alternative versions of Protestantism, both took their lead from Augustine. Both preached that human beings are bad, or at least guilty until told otherwise; but still, the source of our in-built guilt, and why our inherited sin is supposed to be so grave, are hard to fathom. I take Christianity very seriously and spend a lot of time in churches and seek out the company of clerics. But I suggest nonetheless that this doctrine, which has played such a large part in the history of Christianity, is very hard to understand. In some interpretations at least, it certainly seems to put humanity – and many have argued that it seems to put God – in a dubious light. Prophecies tend to be self-fulfiling and maybe life is as fraught as it is in part because our forebears traditionally took such a dim view of themselves and of each other.

Our view of wild creatures and of Earthly life in general has been similarly jaundiced. Nature through most of history was generally seen to be threatening. Even in the pastoral passages of Shakespeare (and very obviously for example in *King Lear),* nature has menace. Shakespeare lived through the preliminary decades of modern science (he was born in the same year as Galileo). He was a seriously knowledgeable naturalist and a keen herbalist, and was clearly aware of early seventeenth century astronomy. Prospero, the magus in *The Tempest,* may properly be seen as a scientist, or at least as a proto-scientist – very possibly based on the then very famous John Dee. Yet Prospero's island, for all his personal enlightenment, is just as magically threatening as the woods of *Midsummer Night's Dream.*

The twin themes, of nature's menace and of deeply embedded human unworthiness, continued through the Renaissance and into the Enlightenment. Nature had to be kept at arm's length, or else decisively tamed. Civilisation, in large part, was a matter of taming it. The Tudors of the sixteenth century favoured knot gardens, knee-high hedges of box and yew intricately interspersed with medicinal herbs, and peaches and apricocks (as Shakespeare was wont to call them) impeccably pruned and pleached. There was always a wall around the garden – partly to create a microclimate but largely to keep the wilderness at bay. The archetypal garden of the seventeenth century is that of Versailles – essentially a knot garden on the scale of an estate. For beyond the sanctuary of the garden wall and perhaps even within it, as Thomas Hobbes famously assured us in the *Leviathan*, published in the wake of the English civil war in 1651: 'The life of man [is] solitary, poor, nasty, brutish, and short'.

Enlightenment intellectuals in the eighteenth century tended to agree with Plato that people at large, meaning most of us, need to be in kept in our place, with philosopher-statesmen in charge. Only a few, of whom Jean-Jacques Rousseau is the best known, spoke up for untamed humanity – he is famed for promoting 'the noble savage', although he never actually used that expression. As for nature: gardens did become more 'wild' in a themed park kind of way and so gave rise to a genre of landscape gardeners of whom the most famous is Lancelot 'Capability' Brown. Brown created extended gardens with grass and woods that imitated nature. As one contemporary commentator put the matter, albeit with some economy of truth, he 'helped nature to become itself'. But Brown's version of nature was highly idealised; and another of his contemporaries remarked that he hoped to die before Brown did, so that he might see Heaven before it was improved.

It wasn't till near the end of the eighteenth century that artists and writers, who in England included such giants as John Constable, Samuel Taylor Coleridge, and William Wordsworth, finally allowed themselves to glory in nature as it really is – positively to relish the wildness. Even then the poets in particular tended to 'romanticise' it, with a small 'r', enhancing wild nature with their own imaginations; and although Constable painted wild skies, very beautifully and accurately, he didn't on the whole paint truly wild landscapes. He preferred his native Suffolk, close to the heart of the agricultural revolution. Nonetheless, the Romantics made a true attempt to engage with nature: not to keep it at bay but to seek it out.

By that time, however, nature was already being brought to heel. The reasons to fear its mystery and power and unpredictability were being chipped away. Explorers and natural sciences had begun to show us that the real world is more extraordinary by far than mere human beings, however imaginative, could envisage – but they were exploring it nonetheless, throwing light in its darkest corners, bringing it within human compass. Outstanding were and are the voyages of Captain James Cook in the 1760s and 1770s – the first with the botanist Joseph Banks on board; and the extraordinary, and extraordinarily conscientious explorations of the saintly Alexander von Humboldt (of whom the young Charles Darwin was in awe). At that time too, or thereabouts, the first of the recognisably modern geologists, in particular the two great Scots, James Hutton and then Charles Lyell, began to show beyond reasonable doubt that the Earth is incomparably more ancient than had been assumed – many millions of years at least; and that the deep past was far more turbulent than had been imagined, or could be imagined. The Flood, described in Genesis, was only a part of what had gone on. (Possibly, or even probably, so some geologists say these days, Noah's flood was the thaw at the end of the last Ice Age, or at least a folk memory of it). The young Darwin devoured the writings of Lyell, too.

From the seventeenth century, too, and through the eighteenth and nineteenth, the new and powerful technologies of the industrial revolution encouraged a shift of consciousness. For although some at least of the Romantics (including Rousseau) railed against the factories and all that went with them, others gloried in them, including English painters such as James Ward and Joseph Wright of Derby – and, later, J.M.W. Turner, who many consider to be the greatest English painter of all. Turner was excited by the drama and the colours of steam, even if he couldn't convey the noise of it, or the sulphurous tang of the burning coal.

Most of all, though, perhaps, industrialisation seemed to fulfil the Enlightenment dream, at once exciting and deeply pernicious, that human beings, if they choose, can be masters of all. Wild nature, for all its magnificence and power, need hold no fears. We needn't exclude nature behind stone walls, and make our own little havens within. If we chose we could simply encompass the lot. Great wild gorges were spanned by great iron bridges, by Thomas Telford and Isambard Kingdom Brunel. In truth, too, most of the Romantic poets weren't quite as wild as they liked to pretend. On the whole they preferred to stay in nice hotels and, at least after the 1840s, to travel by train. John Clare of Northamptonshire in the English midlands truly did engage with nature – he was a farm labourer before he was a poet – and was apt to take John Keats to task for his inaccuracies. (He probably didn't say 'None of yer fancy ways!' but the sentiment was there). All in all, it seems as if artists did not venture to embrace wild nature until they felt pretty sure that they could bring it under control. They weren't so intrepid as our ancient ancestors who truly ventured into the unknown and encountered mysteries and gods and demons and sacred places wherever they went.

But the insights of science, and the power of the new technologies, brought terrors of their own. In particular, between 1798 and 1826 the English economist-cleric Thomas Robert Malthus, known to his friends as 'Bob', published six editions of *An Essay on the Principle of Population* – in which he assured us all that in effect, we're doomed. His message, stated baldly, is that the human population tends to increase exponentially, which means by a certain *percentage* each year, which means that the growth becomes faster and faster and faster – like compound interest. But our food supply can be increased only by a fixed amount each year, at best; and sooner or later must reach a limit. So, he said, sooner or later human populations are bound to outbreed their resources. Or rather, as befits a man who had won prizes for rhetoric at Cambridge, he wrote:

> The power of population is so superior to the power of the earth to produce subsistence for man, that premature death must in some shape or other visit the human race. The vices of mankind are active and able ministers of depopulation. They are the precursors in the great army of destruction, and often finish the dreadful work themselves. But should they fail in this war of extermination, sickly seasons, epidemics, pestilence, and plague advance in terrific array, and sweep off their thousands and tens of thousands. Should success still be incomplete, gigantic inevitable famine stalks in the rear, and with one mighty blow levels the population with the food of the world.[3]

Powerful stuff. We know from his notebooks that Darwin read Malthus as a young man, and many people in high places remain Malthusian to this day, convinced that rising population is the prime source of the world's ills, dwarfing all other problems. Indeed the arithmetic can suggest this. Modern concern with population was perhaps at its height in the 1960s when human numbers were growing at around two per cent per year. This you might suppose would double the population in 50 years (50 x 2 = 100). But because of the compound interest effect, a population increasing at two per cent per year would double in forty years. In 100 years – which, we could say, is within a human lifetime – a population that continued to grow at two per cent would increase about six times. Two per cent per year might not seem too frightening. A six-fold increase in total numbers within the plausible lifetime of one person, certainly does.

For all those people need to eat. They need other resources too, of course, but they need food every day. When Malthus was writing, Britain had already seen much of the agricultural revolution of the seventeenth and eighteenth centuries, with extensive drainage, new methods of fertilising, new and more complex rotations, new ingenious machines for cultivation and sowing, and new breeds of livestock and varieties of crops (for plant and livestock breeding were highly sophisticated by the end of the eighteenth century). Yet, said Malthus, the increase in the output of food must soon reach a limit. Human numbers, increasing exponentially, must soon exceed those limits. Therefore, he said, a crash is inevitable.

In short, through the end of the eighteenth and into the start of the nineteenth centuries ancient ideas lingered on – including the ancient pessimism, including a very the dim view of nature and of human beings; while a whole range of new insights and possibilities – and new fears – came on board. In Britain, human numbers rose spectacularly as agricultural output increased, and the country imported goods of all kinds from around the world to swell its new industries, sometimes at huge cost to the exporting countries, including India and much of Africa. To add to the upheaval there were revolutions in America in 1776 against the British, and in France in 1789 against the aristocracy, after which Napoleon Bonaparte decided to take over the known world in the style of ancient Rome. The Napoleonic wars came to an end in June 1815 with the Battle of Waterloo – but a couple of months earlier, in April, a volcano called Tambora exploded on the Indonesian island of Sumbawa and filled the upper atmosphere with ash which, in effect, blotted out the sun for the rest of that year and the whole of 1816. J.M.W. Turner produced some of his most spectacular sunsets in the wake of it, which is a bonus. Mary Shelley wrote *Frankenstein* while she waited for the rain to stop during a holiday to the Alps, which is a bonus too. But also, which definitely is not, crops failed worldwide. In June 1816 there were frosts in Connecticut and snow in New York. The New Englanders spoke of 'Eighteen-sixteen and froze to death'. Of course, there were food shortages as harvests failed. Add all that to the general political upheaval, the explorations and the rise of science and of big technologies including steam power, the urbanisation and the rapid rise of human numbers, and we have turbulence indeed, and a great deal of suffering.

As the nineteenth century wore on the technologies grew mightier, Britain's empire expanded and consolidated, while the cities swelled and although many became richer and more comfortable, life for the world at large became ever more fraught, complicated, and uncertain. In the 1830s Alfred Tennyson, in his *In Memoriam,* wrote of 'nature red in tooth and claw' – and so indeed it must have seemed. In the cities, Malthus's dire predictions seemed already to be coming true. When Ebenezer Scrooge was asked to support poor orphans in Charles Dickens's *A Christmas Carol*, published in 1843, he observed that their deaths 'will only lower the surplus population'. A couple of years later, in 1845, came the start of the Irish Potato Famine. Over the next few years a million people died of starvation and another million emigrated to America and elsewhere – reducing the population of Ireland by about a quarter. Western Scotland and parts of western Europe suffered as well.

In the early nineteenth century, too, the world as a whole was still very religious and Christianity in the west was still a huge force – probably as great as ever. Yet at the same time, for all kinds of reasons that surely included the legacy of the eighteenth century Enlightenment and the rise of geological science and the spectacle of general suffering worldwide, for reasons both natural and all too human, orthodox Christian theology began to lose its grip. This is reflected in literature through most of the nineteenth century, from Fyodor Dostoyevsky to George Eliot to Thomas Hardy – and to many, the loss of faith was a matter for deep regret. Matthew Arnold most memorably caught the general mood in *Dover Beach* in 1867:

The Sea of Faith

Was once, too, at the full, and round earth's shore
Lay like the folds of a bright girdle furled.
But now I only hear
Its melancholy, long, withdrawing roar,
Retreating, to the breath
Of the night-wind, down the vast edges drear
And naked shingles of the world.

In short, the years from about 1770 to the mid-nineteenth century were among the liveliest and most turbulent in all human history. There was a complete shift in just about everything, much of it painful, offering no apparent reason at all to revise the ancient opinion of nature (basically harsh) or of human nature (basically feckless, at best); and all that, or so it seemed increasingly to many, without the comforting arm of God.

This was the world that Charles Darwin grew up in.

Darwin and *The Origin of Species*

Charles Darwin was born in 1809 in Shrewsbury, in the west midlands of England, the fifth child out of six of Robert Darwin, physician and financier, and grandson of the eighteenth century polymathic physician, scientist, and poet, Erasmus Darwin. Charles's mother, Susannah, was the eldest daughter of Josiah Wedgwood, the great industrial potter and a significant intellectual. In 1839 Charles married his cousin, Emma Wedgwood, and they went on to have nine children. In short, Darwin's life from our present vantage point seems enviable indeed: wealthy, enlightened, scholarly, and firmly rooted in a large and talented family.

He attended Shrewsbury School – needless to say, he is their most famous alumnus – and in 1825, he went up to Edinburgh University to read medicine, in the Darwin family tradition. But he found he didn't like it, although he did meet some good people, including the outstanding naturalist Robert Grant. So in 1828 Charles upped sticks for Cambridge to read for the clergy but again he failed to settle. So he switched again, to natural sciences, and although he left in 1831 with a BA degree but without taking Honours, he showed great promise as a naturalist. So it was that in 1831, when Captain Robert Fitzroy of the Royal Navy was planning a world exploration on HMS *Beagle,* and wanted a gentleman-naturalist-companion, Charles's old Cambridge mentor, the saintly botanist John Stevens Henslow, recommended him for the job. He got it. The *Beagle* set sail from Devonport two days after Christmas 1831, with Darwin, who later proved to be the world's worst sailor, on board. They visited South America, the Galapagos Islands, Tahiti, New Zealand, Australia, and Mauritius – Darwin spent more time on land than at sea – and returned in 1838. Such a trip would change anyone's life. It transformed young Charles from a keen naturalist (although Henslow had said he was not quite 'finished' when he began) into a biologist whose insights have changed the world.

Darwin's great idea

In the forty-odd years after the voyage of the *Beagle* (he lived until 1882), Darwin provided many fine insights into barnacles, coral reefs, earthworms, orchids, climbing plants, plant breeding, and more besides. But the idea that changed all biology and indeed the entire western worldview is that of evolution. In *The Origin of Species* of 1859 he presented overwhelming evidence from many different angles that evolution is indeed a bedrock principle of life. Creatures were not created ready-made in their present form, or even half-made, as prototypes, but evolved over aeons out of ancestral lineages that stretch back to the origins of life itself.

More than that: he suggested a plausible mechanism of evolution – that of natural selection, in uneasy combination with sexual selection (which produces many strange features such as the peacock's tail that clearly get in the way of day-to-day survival). Natural selection also showed, more or less in passing, how it is that almost all creatures are so well adapted to their environments – not only to the physical conditions, but to the creatures they are surrounded with. For example, way back in the 1833 Darwin was asking why it is that dung beetles don't exist in places where there are no large mammals to provide the dung – 'a very beautiful fact, as showing a connection in the creating of animals as widely apart as Mammalia and Insects'.[4] In short: a Creator could not simply create each herb and beastie individually. He would have to design entire ecosystems all at once (although the term 'ecosystem' did not of course exist in the 1830s).

Crucially, too, Darwin suggested – in stark opposition to the general, Platonic, gut feeling of his day – that species could change over time. They could and did 'transmute' to form new species. Indeed – a wonderfully unifying idea – he suggested that all Earthly creatures have descended from a single common ancestor. For good measure, and on a considerable point of detail, he suggested that human beings too are evolved; not created separately on the sixth day as suggested in Genesis (although he didn't expand publicly on this line of thought until 1871 in *The Descent of Man*).

Many since have taken these ideas to reinforce their own atheism, but this is in truth is a hi-jacking. Darwin did at one point, perhaps at various points, profess to have lost his faith (as so many Victorian intellectuals did). But the modern atheist fundamentalists are wrong to invoke him as a natural ally, just as religious fundamentalists are wrong to condemn his ideas out of hand. Many fine modern biologists are deeply religious; and as we will see, a great many clerics had no trouble at all with evolution even in Darwin's own day. Indeed, as I will argue later, a religion that embraces the idea of evolution on the one hand and of transcendence on the other would be very powerful and very much to the point. In fact it could be just what the world needs.

Darwin was not of course the first to suggest that Earthly creatures have evolved from more primitive ancestors. In the eighteenth century Darwin's own grandfather, Erasmus, was thinking along evolutionary lines. Erasmus's near contemporary, the geologist James Hutton, mentioned earlier, was clearly homing on the specific idea of natural selection. For example he said in his *Investigation of the Principles of Knowledge* (1794, volume 2), that since dogs survive by 'swiftness of foot and quickness of sight' then it follows that:

> ... the most defective in respect of those necessary qualities, would be the most subject to perish, and that those

> who employed them in greatest perfection ... would be those who would remain, to preserve themselves, and to continue the race.

More conspicuously, the great French biologist Jean-Baptiste Lamarck expounded evolution at the end of the eighteenth century and the start of the nineteenth – but he thought that change could come about by 'inheritance of acquired characteristics' rather than by natural selection. In 1844 the Scottish publisher, geologist and general polymath Robert Chambers published – anonymously – his *Vestiges of the Natural History of Creation* in which in particular he discussed the key idea of the 'transmutation of species' – the notion that species may change over time and so give rise to new species. Chambers' book (still anonymous) became an international bestseller, and as Darwin later acknowledged, it helped to pave the way for his own ideas. Even earlier than Chambers, in 1831, the Scottish landowner and naval engineer Patrick Matthew outlined the key features of natural selection, the core of Darwin's own theory – but, curiously, he presented his evolutionary ideas in an appendix to a book on wood for ships (*On Naval Timber and Arboriculture*) where, unsurprisingly, as Darwin himself pointed out, it was overlooked.

Most strikingly of all, in 1858, while collecting in the East Indies, the naturalist and explorer Alfred Russel Wallace wrote a convincing and brilliant account not only of evolution in general but of natural selection in particular (although he didn't coin the expression 'natural selection'). He sent his treatise to Darwin for his comments; and it was this that jolted Darwin into publishing his own ideas. Darwin told Charles Lyell that Wallace seemed likely to beat him to the post – which he rightly felt was galling since he, Darwin, had been pondering the idea for some decades. So as to be fair both to Darwin and Wallace, Lyell and the botanist Joseph Hooker arranged for their papers to be read one after the other at a meeting of the Linnean Society of London, which is still going strong and then was very influential. So the papers were duly read, in 1858, by the Society secretary John

Joseph Bennett, in a session entitled 'On the tendency of species to form varieties, and on the perpetuation of varieties and species by means of natural selection'.

In their initial letter of introduction, Lyell and Hooker referred to Darwin and Wallace as 'the two indefatigable naturalists' – although by the time of the meeting Darwin was far from indefatigable and preferred to stay at home, while Wallace, who truly was indefatigable, was still away collecting. But perhaps it was just well that they stayed away. The paper went down like a lead balloon, with many of the audience seeming 'fatigued'. In his annual report for 1858 the Linnean Society President Thomas Bell famously opined that 'The year which has passed has not, indeed, been marked by any of those striking discoveries which at once revolutionise, so to speak, the department of science on which they bear'. In its enormity this compares with A.A. Michelson's comment towards the end of the nineteenth century that:

> The more important fundamental laws and facts of physical science have all been discovered, and these are now so firmly established that the possibility of their ever being supplanted in consequence of new discoveries is exceedingly remote.[5]

Michelson made this comment shortly before Einstein first announced relativity and Max Planck set physics on the road to quantum theory.

Nonetheless, some have said that Darwin was a Johnny-come-lately: that by the time he published *Origin*, evolution was already 'in the air' if not more or less established, thanks to Robert Chambers and a few others. But it wasn't. Despite Chambers *et al*, special Creation was still the prevailing idea: that each creature had been moulded individually. Darwin's detractors also suggest that he did not properly acknowledge earlier writers in *Origin*, and that in this he was remiss. But he did acknowledge the ones he knew about (and later apologised up to a point to Patrick Matthew, when he was told about him).

In reality, though, Darwin was one of the first on the scene. His notebooks show that as early as July 1837 he was pondering the transmutation of species – and in 1844 (the year of *Vestiges*) he told his friend Joseph Hooker that he believed that transmutation is indeed a fact of life. Transmutation of species is a huge idea. Lamarck, for example, thought of evolution – but he did not suppose that lineages of creatures could branch to produce several different lineages. He seemed to suggest instead that each species had appeared separately in the deep past and had then changed over time – but always staying within its prescribed evolutionary tramlines. Like Plato, Lamarck seemed to conceive each species as an ideal – as some would say, an idea in the mind of God. The notion that lineages of creatures could fork and branch to give rise to many different tracks is far more radical than the mere idea that any one lineage may change over time – not least because it seems to suggest that God could change his mind, and so to offend Plato's notion that species are 'ideals'. So from the outset, Darwin was pushing the boundaries. By 1844 too, he had already completed a 231-page essay on evolution by means of natural selection.

Yet Darwin did not immediately publish. Evidently he realised how monumental his ideas really were, and how they might offend the sensibilities of people he respected, including his old teacher, the devout Reverend Henslow. Or perhaps, not being steeped in neo-Darwinian *angst* as scientists are today, he did not feel the breath of competition and simply saw no reason to rush. Instead, in 1846, he embarked on an exhaustive study of barnacles. He finally began what was intended to be the definitive work on evolution in 1856, but with no apparent deadline in mind. *Origin* was a rush job. He started work on it only after Wallace's letter arrived in 1858 and although his master-work ran to 500 pages when it first appeared on November 24, 1859, it was intended only as a summary of his thoughts so far. But published it was, at last, and with a comprehensive and resounding title: *On the Origin of Species by Means of Natural Selection, or the Preservation of Favoured Races in the Struggle for Life*. Darwin did not provide an exhaustive bibliography in *Origin* because he never intended it to be the definitive work. The definitive work that he began in 1856 was never completed. This is often the way with definitive works.

The basic ideas have been described many times but they are so crucial to modern western thinking that it's worth running through them to see how they arose and where they lead us. The thesis is of course brilliant yet with an unfortunate emphasis that has proved extremely damaging.

Darwinian evolution in a nutshell

The idea of natural selection is based on a shortlist of irrefutable observations and a simple chain of logic – all so neat that Darwin's great friend and supporter the sometimes fierce Thomas Henry Huxley (known to Darwin's somewhat awe-struck children as Uncle Henry) said that those who had not thought of it, which is most people, should think themselves very foolish.

First, we may simply observe that living creatures invariably, at some stage, reproduce. They have offspring. Replication is not vital to life, however. Individual organisms will often live longer if they *don't* produce offspring. But time and chance ensure that any one thing, whether it's a temple or a mountain or a living creature, will fall apart sooner or later. Reproduction doesn't keep the parent alive, and often hastens the parents' end. But offspring can ensure that the lineage continues. Living creatures that don't reproduce are a dead end (just as temples and mountains are dead ends) but those that do reproduce are in with a chance and the fact that life is now so abundant shows that the general strategy – replicate before you die! – works. Or at least it has worked well enough for the past 3.8 billion years.

But, said Darwin, *all* creatures are always prone to produce more descendants than their environment, or indeed the whole world, can possibly sustain. In the pure spirit of Malthus he writes:

> There is no exception to the rule that every organic being naturally increases at so high a rate, that if not destroyed, the earth would soon be covered by the progeny of a single pair.[6]

This is just as true of creatures that breed slowly as of those that have thousands of offspring: 'Even slow-breeding man has doubled in twenty-five years'. Indeed

> The elephant is reckoned to be the slowest breeding of all known animals and I have taken some pains to estimate its probable minimum rate of natural increase ... [yet] at the end of the fifth century there would be alive fifteen million elephants, descended from the first pair. *(Ibid.* p.587)

In fact, he says:

> We may confidently assert, that all plants and animals are tending to increase at a geometrical ratio, that all would most rapidly stock every station in which they could any how exist, and that the geometrical tendency must be checked by destruction at some period of life. (p.589)

Now the first big crunch:

> Hence, as more individuals are produced than can possibly survive, there must in every case be a struggle for existence, either one individual with another of the same species, or with individuals of distinct species, or with the physical conditions of life. It is the doctrine of Malthus applied with manifold force to the whole animal and vegetable kingdoms; for in this case there can be no artificial increase in food, and no prudential restraint from marriage. (p.597)

He emphasises this chill expression – 'struggle for existence', or 'struggle for life' – even if he doesn't much care for the idea of it:

> Nothing is easier to admit in words the truth of the universal struggle for life, or more difficult – at least I have found it so – than constantly to bear this conclusion in mind. Yet unless it be thoroughly engrained in the mind, I am convinced that the whole economy of nature, with every fact on distribution, rarity, abundance, extinction, and variation will be dimly seen or quite misunderstood. (p.586)

It's obvious too – a second strut in Darwin's argument – that like begets like. Horses give birth to foals and spiders give rise to baby spiders. Only in mythology (Greek, Chinese, what you will) do people and animals produce offspring that are not of their kind.

But, said Darwin, no two offspring of creatures that reproduce sexually are ever identical: 'Amongst organic beings in a state of nature there is some individual variability' (p.585). He doesn't set out to prove this but, he says, 'I am not aware that this has ever been disputed'. (Of course if creatures reproduce by cloning – a form of asexual reproduction – as many do in nature, then the offspring will be more or less identical. Even in clones, though, there is likely to be *some* variation, for reasons we will touch upon later).

Now another key line of thought: the notion that species are not immutable: that one kind may indeed 'transmute' into another; that one lineage can give rise to several. This notion first came to him in the late 1830s, in the wake of his voyage on the *Beagle.* For when Darwin set sail in 1831 he did not suppose that any one species could alter enough to 'transmute' into a different species. No biologist at that time thought such a thing was possible. He assumed (although it's a very strange assumption when you think about it) that new species arise in different 'centres of creation'. The idea of transmutation from species to species began to dawn during or not long after the stopover on the Galapagos Islands (I'm told there are fifteen biggish ones and three small ones), in 1835. There Darwin found, although it didn't strike him immediately, that different islands had their own species of giant tortoise. He also saw that different islands had different species of mockingbird – and he perceived that those mockingbirds resembled those of mainland America, 600 miles to the East, although they weren't the same as the mainland kinds.

But most famously, Darwin and his companions and assistants, including Captain Fitzroy himself (who was a fine thinker in his own right and a keen amateur scientist) also noted and collected a range of what they called 'finches'. Back in England in 1837 the ornithologist John Gould took a long look at the finches and showed that they belonged to twelve different species (fourteen are now known) – including one that Darwin had called a 'wren'. But they weren't true finches at all, said Gould. They belonged to a quite different family, and he couldn't say who they were related to. (But they're still called 'Darwin's finches' nonetheless).

The Galapagos finches all have rather dull plumage and Darwin, who was not an ornithologist, didn't take proper notes. He didn't even note where they were all found – but fortunately some of his shipmates did and with the help of Gould, he was able to see that different islands tended to have their own species of finch, just as with the tortoises and mockingbirds. Also, and even more strikingly, the different 'finches' took many different forms; some quite big and some very small; some with heavy bills for crushing big seeds like European bullfinches and haw-finches, and some with needle-like bills like European warblers, for prising out small insects.

In all, Gould found that Darwin and his companions had collected 26 different species of birds – 'finches', mockingbirds, and a few others – of which 25 were new to science. Clearly they were allied to birds of the South American mainland but equally clearly, they were different; unique to the Galapagos.

The most economical explanation of all this, Darwin began to realise, is that the ancestors of the tortoises, the mockingbirds, and the finches had arrived some time in the deep past from the American mainland – the birds flying in or blown off course; the tortoises presumably 'rafting' on floating vegetation. Perhaps, or probably, only one of a few of each original kind of bird or tortoise had made it to the islands. Then, in each case, the pioneers had evolved to form the present 'suites' of different but related species that we see today. The finches in particular had diversified to fill niches – big seed-eaters, little seed-eaters, insect-eaters – that on continents are already filled by many other birds; but which, on the volcanic Galapagos Islands, thrust up from the ocean floor, were vacant. The notion of species transmutation clearly dawned on Darwin not long after his conversations with Gould and he recorded it formally and publicly as early as 1845, in the second edition of *The Voyage of the Beagle*:

> Seeing this gradation and diversity of structure in one small, intimately related group of birds, one might really fancy that from an original paucity of birds in this archipelago, one species had been taken and modified for different ends.[7]

In *Origin* Darwin integrated his notion of species transmutation with observations he had been making in domestic breeding – for he was absorbed by biology in all its manifestations, domestic and homely as well as the most exotic. He consorted with breeders of fancy pigeons and goldfish, studied gardening magazines, and pondered the mysteries of crop improvement which in his day, with the agricultural revolution still in full swing (it has never really stopped) was a hot topic. Crop and livestock 'improvement' depend on selection – 'artificial selection'. The breeder starts with the pigeons with the most absurdly pouting chest or the fish with the most voluminous tail or the bull with the squarest hindquarters or the wheat with the fullest grains, and crosses them with others with the same or complementary qualities and then breeds from the offspring that express most strongly the quality that's desired, and so on and so on. In a short time, artificial selection produces a huge range of variations which may differ enormously from the ancestral type. Thus huskies look roughly like wolves which are the ancestors of all domestic dogs and it's easy to see how you might derive a German Shepherd from a wolf. But a St Bernard? A Pekinese? A Chihuahua? It hardly seems possible – and yet it clearly happens. Yet most naturalists of the time – including Wallace – thought that no amount of selection could produce brand new *species*. St Bernards and Chihuahuas differ far more in appearance than, say, Ravens differ from Rooks. Yet St Bernards and Chihuahuas still remained resolutely within the same species, *Canis familiaris* – while Ravens and Rooks, superficially similar, are *qualitatively* different; and no amount of selection could never produce a qualitative shift. That was the dogma.

But Darwin said – why not? The difference between the variety on the one hand and the species on the other is only one of degree. It would take longer to produce a new species than to produce a new variety – but that's the only real difference. The point, of course, is that there is a reproductive barrier between Ravens and Rooks that does not exist between different breeds of dog (apart, of course, from obstacles raised by differences in anatomy). But given time, reproductive barriers could surely arise between different races which then would qualify as different species:

> How fleeting are the wishes and efforts of man! How short his time! And consequently how poor will his products be, compared with those accumulated by nature during whole geological periods. Can we wonder, then, that nature's productions ... should plainly bear the stamp of far higher workmanship? (p.162)

Darwin then put all these ideas together: that all creatures have a tendency to out-breed their resources; so there is a struggle for life; and that nature will select the individuals best suited to the prevailing conditions, just as breeders of crops and livestock select whatever takes their fancy. As he put the matter himself:

> Owing to this struggle for life, any variation, however slight and from whatever cause proceeding, if it be in any degree profitable to an individual of any species ... will tend to the preservation of that individual, and will generally be inherited by its offspring ... I have called this principle, by which slight variation, if useful, is preserved, by the term of Natural Selection, in order to mark its relation to man's power of selection ... But Natural Selection ... is a power incessantly ready for action, as is as immeasurably superior to man's feeble efforts, as the works of Nature are to those of Art. (p.586)

Very significant, I think, is the humility that Darwin expresses in this last sentence: that nature is 'immeasurably superior to man's feeble efforts'. Elsewhere he tells us:

> It may be said that natural selection is daily and hourly scrutinising, throughout the world, every variation, even the slightest; rejecting that which is bad, preserving and adding up all that is good; silently and insensibly working, whenever and wherever opportunity offers, at the improvement of each organic being in relation to its organic and inorganic conditions of life. We see nothing of these slow changes in progress, until the hand of time has marked the long lapse of ages ... we only see that the forms of life are now different from what they formerly were. (p.162)

Passages like this prompt me to observe that Darwin can properly be called a Romantic – far closer to Coleridge than to Dawkins. He stresses, though, that we, mere human beings, just don't *know* what features nature will choose to favour not least because the 'modifications [that] are accumulated by natural selection for the good of the being will cause other modifications, often of the most unexpected nature'. Sometimes, though, our own feeble efforts with artificial breeding offer clues to what might be going on. Thus on page 161:

> In plants the down on the fruit and the colour of the flesh are considered by botanists as characters of the most trifling importance: yet we hear from an excellent horticulturalist, Downing, that in the United States smooth-skinned fruits suffer far more from a beetle, a curculio, than those with down; that purple plums suffer far more from a certain disease than yellow plums; whereas another disease attacks yellow-fleshed peaches far more than those with other coloured flesh. If, with the aids of art, these slight differences make a great difference in cultivating the several varieties, assuredly, in a state of nature, where the trees would have to struggle with other trees and with a host of enemies, such differences would effectually settle which variety, whether smooth or downy, a yellow- or purple-fleshed fruit, should succeed.

We can easily see from all this how the mechanism of natural selection explains at a stroke the perennial question – a question first posed by theologians – of how it is that all Earth's creatures are so well adapted to their circumstances. Nature, unlike capricious human beings, does not select the birds that are most likely to win rosettes, or cattle that fetch the highest price once slaughtered. Perforce it selects the individuals that survive the best in any one set of circumstances – which of course are the ones that are best adapted; most suited to those circumstances. In fact, in the 1860s, the philosopher Herbert Spencer summarised natural selection as 'survival of the fittest' – an expression that Darwin himself adopted in later editions of *The Origin.*

Loud and clear, though, in Darwin's account of evolution by natural selection, is the absolute importance of competition: with the prevailing conditions, and particularly climate; with other individuals of the same species; and with creatures of different species. If there was no competition there would be no evolutionary change. After all, if all individuals survived in any one set of circumstances irrespective of their aptitude, there would be no selection. All the various types would have an equal chance of surviving. Thus he gives the impression that life is one long punch-up – 'red in tooth and claw' just as Tennyson had suggested. By an etymological accident, Herbert Spencer's choice of words – 'survival of the fittest' – reinforces this impression. After all, 'fit' in the nineteenth century commonly meant 'apt', as in 'It is fitting that a gentleman should remove his hat when introduced to a lady'; or as in the modern 'not fit for purpose'. But 'fit' also in modern English means strong, healthy, able to do a great many press-ups. So it is that so many modern people clearly believe that Darwin told us that competition is the key to change and hence to progress, and that the best competitors are the toughest. I reckon that Darwin did over-emphasise the competitiveness of life. But Spencer made it worse, and generations of Darwin's disciples since have turned his wonderful insight into something rather ugly, and most unfortunate.

But I will come to that. First, in Chapter 2, we should see how Darwin's great idea was combined with other great ideas – the genetics of Gregor Mendel, molecular biology, and the maths of game theory – to create the modern form of 'neo-Darwinism'. Neo-Darwinism, like Darwin's own original theory, is a superb piece of science – but also, like Darwin's original theory, it has taken some pernicious turns.

The evolutionary biologist and geneticist Theodosius Dobzhansky famously commented in 1973 – it was the title of an essay – that 'nothing makes sense in biology except in light of evolution'. This surely is true – and Darwin should surely be seen as the greatest of all biologists simply because he presented the idea of it so cogently and comprehensively. Yet even more importantly, he was a true and great naturalist. Through all his life – a long life by the standards of his day – he was besotted by nature, and humble in the face of it. Darwin is commonly presented to us as if he was simply continuing the Enlightenment – steeped in the notion that everything can be explained or explained away in mechanistic terms. But in truth he was right at the heart of Victorian Romanticism, both of its lyricism and of its recognition of life's uncertainties. He recognised, as theologians take for granted and as all wise people should acknowledge, that in the end all is mystery; in the end, life and the universe are beyond our ken. This, not the desire for literary flourish, is the source of his own lyricism; and this, of all the lessons that science can teach us, is the greatest of all, albeit so often put aside in this age of anthropocentric, gung-ho materialism.

But although almost all biologists these days accept evolution as a fact of life, some question the particular importance of natural selection – which was at the core of Darwin's great idea. For my part, I am sure as many biologists are that natural selection is indeed a key player (for as John Maynard Smith was wont to say, nothing else can adequately explain in non-creationist terms the way that adaptations are so finely tuned; how exquisite are the interactions between different creatures, and different parts of the same creature). But we surely should question (as many biologists do these days) the absolute importance of competition in driving natural selection. Is life really as competitive as Darwin – following Malthus, or indeed Tennyson – supposed it to be? We will look at this, in various contexts, in Chapters 2 to 6.

2. The Path To The Selfish Gene

As happens with all great ideas – and should happen, or else the ideas die – a great many thinkers from many different disciplines, scientific and non-scientific, picked up the baton that Darwin passed on to us and ran with it. To a great extent their running has been fruitful. Darwin's idea of evolution (with or without the specific notion of natural selection) has triggered some of the most intriguing discussions of the past century and a half, and indeed of all of time, and has provided biology with the kind of all-embracing narrative that implicitly and sometimes explicitly is the goal of all science. Dobzhansky was more or less right: 'Nothing in Biology Makes Sense Except in the Light of Evolution'.

But the modern interpretation of Darwin's ideas, at least as widely understood and even more widely promulgated, has also taken some most unfortunate turns. In particular, Darwin's successors have emphasised the feature of his thinking that is the least attractive – which he himself found least attractive! – and can be damagingly misleading: that life is driven above all by competition; that it's one long punch-up from conception to the compost heap. His successors, too, unbidden and entirely unjustified, have used his name to reinforce their own bleak view of life: one of extreme materialism and fundamentalist atheism (although many good Darwinians are devout, including the Russian Orthodox Dobzhansky). Then again, evident throughout *Origin of Species* and indeed through all his writing is Darwin's personal humility in the face of nature. So he coined the expression 'natural selection' by analogy with the artificial selection through which breeders have produced so many fancy varieties of dogs and pigeons and goldfish. But, as we have already seen, he said that:

> Natural Selection ... is as immeasurably superior to man's feeble efforts, as the works of Nature are to those of Art.

This is in marked contrast to the attitude of gardeners from earlier centuries (including Capability Brown), who took it to be self-evident that nature had to be improved upon. It contrasts even more starkly with the arrogance of the modern-day commercial 'genetic engineers' who seek to tailor animals and plants to suit the commercial whims of the day, and promise a day when they will re-build life from scratch, and even in moments of extreme hubris are happy to speak of 'playing God' – and yet see themselves as natural successors of Darwin. In truth they simply re-cast Darwin in their own image – ultra-'rationalist'; through-and-through materialist.

I see Darwin as one of the great nineteenth century Romantics; the grim, Tennysonian side of his writing – 'nature red in tooth and claw' – was pure gothic (just as it was in Tennyson). I am sure that Dean Farrar hit the nail on the head in the funeral speech that he delivered for Darwin in Westminster Abbey in April 1882:

> This man, on whom for years bigotry and ignorance poured out their scorn, has been called a materialist. I do not see in all his writings one trace of materialism. I read in every line the healthy, noble, well-balanced wonder of a spirit profoundly reverent, kindled into deepest admiration for the works of God.[8]

So let us look, first, at how Darwin's ideas have been expanded and in some ways made even grander in the century and a bit since his death – but have also been transformed and often mangled, to the point where we're now presented with a set of notions that are in large measure repellent; and indeed, in their extrapolated form, are threatening to kill us all.

The road to neo-Darwinism and 'the selfish gene'

Darwin was the greatest of all biologists but Gregor Mendel, the founder of modern genetics, just thirteen years younger than Darwin, at least in some ways ran him close. Like Darwin, Mendel was a great experimentalist; and, also like Darwin, he was happy to carry out experiments with zero equipment and at practically zero expense. Thus while Darwin offered momentous insights into ecology from observations of earthworms in his own garden in Kent, Mendel changed the world with his experiments on garden peas, runner beans, antirrhinums and other homely cultivars in the garden of St Thomas's monastery in Brno, then in Moravia and now in the Czech Republic, where he was a monk and, later, became the abbot.

Mendel (like Darwin) was wonderfully thorough – for, as he commented in his seminal paper of 1866 ('Experiments in Plant Hybridisation') 'a generally applicable law ... can only be arrived at when we shall have before us the results of detailed experiments made on plants belonging to the most diverse orders'. His best-known trials were with garden peas (*Pisum*) – which he tells us he began way back in 1854 with '34 more or less distinct varieties of Peas ... obtained from several seedsmen' which were then '... subjected to two years' trial'. Then, 'for fertilisation, 22 of these were selected and cultivated during the whole period of the experiments'.

His method was to cross – 'hybridise' – different strains of peas that had different characteristics (characters) and to see how those characters manifested in the hybrid offspring. In particular:

> The various forms of Peas selected for crossing showed differences in length and colour of the stem; in the size and form of the leaves; in the position, colour, size of the flowers; in the length of the flower stalk; in the colour, form, and size of the pods; in the form and size of the seeds; and in the colour of the seed coats and of the albumen [by which he means the seed leaves, or cotyledons].[9]

He noted though – a key point, as we will see later – that not all characters are easy to quantify:

> Some of the characters noted do not permit of a sharp and certain separation, since the difference is of a 'more or less' nature, which is often difficult to define. Such characters could not be utilised for the separate experiments; these could only be applied to characters which stand out clearly and definitely in the plants.

Just from these trials in cross-breeding – in effect a quantified exercise in gardening – he outlined the key features of genetics, at least as it was understood until well into the twentieth century. First he showed beyond all doubt (as in truth professional breeders already knew, at least intuitively) that characters are passed on from generation to generation in the form of discrete 'factors' (*anlagen*) which later were called genes; and these genes can be passed on separately and in more or less any combination. So the units of inheritance are not like inks, which flow together and mix (as Darwin at one point supposed was the case). Thus he solved one of Darwin's great problems: how it is that some characters are passed on unchanged from generation to generation while others (like blue eyes in human families) may disappear in one generation and then reappear in a later one. For, wrote Mendel:

> ... those characters which are transmitted entire, or almost unchanged in the hybridisation, and therefore in themselves constitute the characters of the hybrid, are termed the *dominant*, and those which become latent in the process *recessive*. The expression 'recessive' has been chosen because the characters thereby designated withdraw or entirely disappear in the hybrids, but nevertheless reappear unchanged in their progeny.

So it is that in peas, for example, the gene that underpins round seeds seems to dominate the gene that produces wrinkled seeds;

and green pods triumph over yellow pods; and so on. Sometimes, however – as in his trials with runner beans (*Phaseolus*) with different coloured flowers – all is not so simple:

> Apart from the fact that from the union of a while and a purple-red colouring a whole series of colours results, from purple to pale violet and white, the circumstance is a striking one that among 31 flowering plants only one received the recessive character of the white flower, while in *Pisum* this occurs on the average in every fourth plant.

The key point here is that many (in fact most!) characters in most creatures are 'polygenic' – underpinned by several or many genes working in combination. Mendel didn't coin the term 'polygenic' but he had clearly worked this out too:

> Even these enigmatic results, however, might probably be explained by the law governing *Pisum* if we might assume that the colour of the flowers and seeds of *Phaseolus multiflorus* is a combination of two or more entirely independent colours, which individually act like any other constant character in the plant.

But of course he is not content to leave it there. In the pure spirit of modernity he writes:

> It would be well worth while to follow up the development of colour in hybrids by similar experiments, since it is probable that in this way we might learn the significance of the extraordinary variety in the colouring of ornamental flowers.

Excellent. Intuitively, at least in retrospect, it seems obvious that we have only to put the genetical musing of Mendel together with Darwin's idea of evolution, to provide a beautifully rounded picture of how life works. After all, Darwin was well aware that his own theory of evolution needed a theory of inheritance to back it up. How was it, after all, that creatures do indeed produce offspring that resemble themselves ('like begets like') and yet those offspring are varied (so that replication in sexually reproducing creatures is never quite precise)? How is it that characters appear and disappear in any one lineage as the generations pass apparently so haphazardly? Darwin's own ideas on the mechanism of inheritance were bizarre (as his friend and critic T.H. Huxley gently but firmly pointed out). But Mendel's ideas, of genes that are sometimes dominant and sometimes recessive and often work in combination, solve most of Darwin's dilemma at a stroke; and if we throw in the idea of mutation, of which more later – that genes are not always replicated accurately, and so produce the occasional wild card – then we have a more or less complete explanation.

Mendel apparently felt that this was so. His own great paper appeared seven years after *Origin of Species* and he knew of Darwin's work. Accordingly, he sent Darwin a copy of his paper. We know he did because it was found on Darwin's desk. We also know that Darwin didn't read it because the pages remained uncut. If Darwin had read Mendel's paper then, keen peruser of gardening magazines that he was, he surely would have realised its significance. Mendel, for his part, although he had good connections in European science (he was a *protégé* and friend of the great Christian Doppler of the Doppler effect, among others) he was also a monk of peasant birth and was apparently too modest for his own good, and did not persist. As it was, *nobody* took any notice of Mendel's work until late in the nineteenth century when it was rediscovered. It was launched on the world in the early twentieth century – when Mendel's working title 'hereditary factor' was replaced with the felicitous and resonant soubriquet 'gene'.

But even then, when Darwin's ideas had been around for decades and Mendel was very definitely in vogue, the penny did not drop – that the two belong together. Indeed, for a time in the early twentieth century, Mendel's ideas were seen to have superseded Darwin's – or at least to have discredited Darwin's particular brainchild of natural selection. It was left to a shortlist of early-to-mid twentieth century biological and mathematical giants to put the two sets of notions together. They included Dobzhansky, as cited above: the Americans Ernst Mayr and Sewall Wright; and the Britons, Julian Huxley, J.B.S. Haldane, and the statistician R.A. Fisher.

Before those giants got seriously stuck in, however, Mendel's and Darwin's ideas seemed incompatible. For the main take-home message of Mendel's work seemed to be that genes either produce one kind of character, or they produce a quite different variant. It is hard to see how there could be a smooth transition over time from one character to another. Yet Darwin stressed that evolutionary change is gradual – little by little, without discontinuities, for aeon after aeon.

Yet Mendel's paper of 1866 seems to solve the problem easily enough. Though some features of peas (like green pods or yellow pods) are either one thing or another with nothing in between, some 'have a "more-or-less" character'. And although the flowers of peas are either definitely white or definitely yellow the offspring of white-flowered and red-flowered runner beans may be various shades of pink. Now we know that most characters of most organisms are 'polygenic' which Mendel himself clearly realised, although he did not use that expression. Once we see that, it's obvious how individual characters, and of course entire beasties, can change ever so little-by-little as the generations pass. So there is no incompatibility.

Only one more idea was needed to complete the picture: that of mutation: the idea that genes may change over time to produce new variants – variants known as 'alleles' – which produce new effects. This idea was pinned down in the 1920s primarily by Hermann Joseph Muller and his colleagues who showed how X-rays can produce genetic variation. Mutations occur naturally, and when they occur in gametes – eggs or sperm – or in the germ cells that give rise to gametes, they become part of the genetic makeup of the offspring. Germ cells do not often mutate – but lineages of organisms over time produce many billions or trillions, or billions of trillions, of germ cells – so rare events soon mount up. Most mutations have no noticeable effect on the offspring and most of the effects that do result are detrimental; but some mutations produce novel effects that can be put to good use. The variety of colours in the flowers of Mendel's peas and runner beans comes about because different individuals carry different variants (alleles) of the genes that initiate the production of pigments (or which stop the production of pigments). Darwin's idea of natural selection requires a source of variation – for if all offspring are the same, then the concept of 'selection' becomes meaningless. Mutation explains how variation can come about.

So by the 1940s the two great sets of ideas, Darwin's theory of evolution by means of natural selection and Mendel's genetics (plus Muller's idea of mutation), were brought together to form what Julian Huxley called 'the modern synthesis' – also known as 'neo-Darwinism'. But neo-Darwinism has undergone some major transformations since the 1940s, so for convenience I will call that pristine version of it 'neo-Darwinism Mark I'. Mark II was born a few years later.

Neo-Darwinism Mark II

One of my own moments of epiphany came in 1959 when I was in the sixth form at Dulwich. Our biology teacher at that time, Colin Stoneman, took the entire form to the Natural Museum in Kensington to see the exhibition mounted by Sir Gavin de Beer to celebrate the centenary of *The Origin of Species* (and, in passing, the 150th anniversary of Darwin's birth). In many ways 1959 was a vintage year for Darwinian thinking – possibly the best time of all. There was a sense of revelation – it seemed that total understanding was now in our grasp. Yet biology still seemed innocent, at least in my memory of it. The sense of wonder had not yet been overtaken by arrogance. For in those days biology was still firmly rooted in natural history; simply in seeing what was out there, and how it all worked. It was motivated not by a desire to control or to exploit the 'resources' of nature but by the 'wonder of spirit' that Dean Farrar identified in Darwin: the spirit that is shared by poets and artists and indeed by a broad lineage of naturalist/ clerics who have contributed so much to the life sciences (and whose ranks were very nearly joined both by Darwin and by Coleridge).

So the centenary exhibition focused on the extraordinary quirks of nature, and yet showed how evolutionary thinking was beginning to explain what was going on. Take for example (just one among many examples displayed at the 1959 exhibition) the case of the swallowtail butterfly known as *Papilio dardanus*, which provides the most brilliant demonstration ever of Batesian mimicry.

The 'Batesian' in 'Batesian minimicry' is named after Henry Bates who accompanied Alfred Russell Wallace to the Amazon in the 1848 and then stayed on for another fourteen years after Wallace became ill and had to come home (although Wallace then took himself off to the East Indies and Malaya, as then they were). It is practised by creatures who themselves may be harmless and make good eating – but are dressed up to resemble creatures that are dangerous. So it is for example that many hover-flies, which are two-winged distant relatives of house-flies and generally are perfectly safe to eat, if you happen to be an insectivore, are striped like wasps, which of course sting, and whose strident colours shout 'Keep off!' The snag is that if such mimics are too successful they are likely to become more common than the dangerous creatures they aspire to imitate. But if the harmless creatures become too common then the warning colours lose their impact – because the predator quickly learns that *most* of the brightly-coloured creatures in its sights aren't actually dangerous at all. They are merely pretending to be. So then the bright colours become an embarrassment – attracting attention with nothing to back it up.

But *Papilio dardanus* – or at least the females – has evolved a wondrous get-out clause. The males all look the same but the females come in at least fourteen different forms – several of which bear an uncanny resemblance to butterflies of different species that are toxic, which predators are careful to avoid. No one *Papilio dardanus* morph is anything like as numerous as its own particular model but all the different morphs belong to the same breeding population (they all mate with the same *Papilio dardanus* males) so the overall population is large. The females achieve such variousness by maintaining key genes in several different allelic forms. The result – a highly various population of the same butterfly, with each variety geared to its own niche – seems too amazing to be true. Yet the conceptually simple mechanisms described by Darwin and Mendel, with help from Muller on mutation, show how it can all come about. That shouldn't take the wonder of it away – because it is proper to marvel at the underlying elegance of the mechanism and ask, 'How could such wonders have come about?' But we will come to that.

This and many other examples, incorporating the notions of Darwin, Mendel, and Muller, belong to what might be called neo-Darwinism Mark I. But by 1959 we already had the first stirrings of neo-Darwinism Mark II. What made all the difference was DNA. For by 1959 the three-dimensional structure of DNA had already been revealed first by the researches of Maurice Wilkins and Rosalind Franklin in London, and then by the insight of Francis Crick and James Watson in Cambridge, who also offered an outline of its *modus operandi*.

Crick and Watson didn't *discover* DNA, as careless accounts often suggest. DNA was first discovered in Switzerland by Friedrich Miescher in 1869 (although he didn't publish until 1871 – and then he called his discovery 'nuclein'). So it just isn't true, as Orson Welles solemnly assured us in *The Third Man*, that Switzerland's sole contribution to modern civilisation is the cuckoo clock. But nobody understood the significance of DNA. Chemically speaking, after all, it was just another organic acid. That it really was important in heredity began to become apparent in the early twentieth century – but even as late as the 1940s, most scientists assumed that genes were made of proteins, because proteins seemed to be the only molecules that were various enough and yet were orderly enough to carry all the details of hereditary information. Erwin Schroedinger – not a biologist to be sure, but a great original thinker – wrote a famous book in 1946 called *What is Life?* which argued this very point: that genes must be proteins.

The coin flipped in the early 1950s when Wilkins and Franklin in particular began to reveal the basic structure of DNA and the penny finally dropped in 1953 when Crick and Watson provided their detailed three-dimensional model – the famous double helix. DNA, it turned out, had just the qualities needed in the stuff of genes – on the one hand neat and tidy and easily replicated, but on the other hand infinitely variable. Almost instantly, it seemed, it became obvious how DNA works its magic. It provides a code which determines the form of proteins. Proteins in living systems are the key players. They form much of the stuff of which cells are made; and, crucially, they function as enzymes – the catalysts that control the whole metabolism. Whoever controls the proteins, in short, runs the show: and, it seemed, DNA was the controller.

So by 1959 the molecular structure and the general role of DNA were known. Already we could see at least in outline how it steers each living creature through its life – different genes are turned on and off at different stages; first to shape the embryo and then the neonate and so on through childhood and maturity until the mistakes build up and we are turned into what Shakespeare's Jacques in *As You Like It* called 'slipper'd pantaloons'. Overall, genetics was augmented by the science of DNA – a science known as 'molecular biology'. Since DNA makes proteins which control the metabolism, molecular biology from the outset also invaded physiology. Since genetics clearly links with evolutionary theory (as shown by the original creators of 'the modern synthesis') it seemed that biology now had a complete narrative. If Darwin plus Mendel (plus Muller) is neo-Darwinism, then Darwin plus Mendel plus the science of DNA might reasonably be called neo-Darwinism Mark II.

All this and a great deal more was on display in London's Natural History Museum in 1959. Heady stuff. I'd wanted to be a vet up till then. After that, it had to be the pure science of biology. For me, though, as I think for a great many professional scientists whom I've got to know since, biology remained a Romantic pursuit, with a big R.

But a great deal has happened since 1959 – in fact one of the most significant shifts of emphasis was already underway (not least in the head of an American biologist, George C. Williams). Since then the ideas have changed – and so, it seems, have attitudes. The Romanticism that once invited us to marvel at nature for its own sake, as an aesthetic and spiritual experience, inviting humility in the face of nature, has to a significant extent been subverted by an ever-increasing urge to use all science as a means to enrichment; to make us (or at least some of us) more comfortable. Sadly, the insights of Darwin have been used to justify an ever more bullish materialism. But we will come to that. Let's look first at the unfolding ideas of the late twentieth century and the early twenty-first.

The shift of emphasis

For Darwin, the struggle for life was between individuals – of the same or of different species – or between each individual and its environment. Natural selection favours the individual creatures who best overcome life's difficulties. But beginning in the late 1950s and through the '60s a group of biologists came on board who argued that in truth, natural selection operates most forcefully not on whole, individual creatures, but on individual *genes*. First among them was the American George C. Williams; then came an Englishman, W.D. ('Bill') Hamilton; and then another Brit, John Maynard Smith, and another American, Robert Trivers. Alas, of the four, only Trivers is still with us. They came at the issues from different angles but their ideas were brought together into one grand thesis – very brilliantly, it should be said – by Richard Dawkins in 1976, in *The Selfish Gene*. But although the book was brilliant, the title was most unfortunate as we will shortly see – misleading and damaging – even if it did help to sell a lot of copies.

Often, it doesn't seem to make much difference whether we suppose that natural selection works most forcefully on whole organisms, or whether it works on their individual genes. Thus we could say – as James Hutton said way back in the eighteenth century – that a dog that runs faster than other dogs would: (a) be able to bring down a greater range of prey (fast deer as well as slow ones), and so win the competition between species; but also (b) get to the prey before the other dogs, and so win the competition with others of its own kind. In Darwinian terms, we could say that natural selection favours individual dogs that are fleet of foot. But after the 1960s it became proper to say that natural selection favoured the *genes* that enable dogs to be swift. So why does that matter?

For a start – and this was Bill Hamilton's great contribution – 'gene level selection' seems to explain at a stroke how it is that some animals (in fact a great many, when you tot them up) from all classes (mammals, birds, fish, insects, etcetera) may act in ways that biologists calls 'altruistic'. That is, they behave in ways that benefit their fellows, or the entire community, even though they themselves seem to suffer by doing so. If all creatures are indeed caught up in a 'struggle for existence', and if their prime and obvious requirement is to stay alive and give birth to others of their own kind, why are so many apparently prepared to risk their own lives, or even lay down their own lives, or give up any hope of reproducing, for the apparent benefit of others – or apparently to serve 'the greater good?'

In an arm-waving way we could argue that many cases of apparent self-sacrifice are, when you boil them down, nothing more than examples of 'enlightened self-interest'. It's often been argued, for example, that this is true of rabbits or of some deer which flash a white tail as they retreat from the hawk or the lynx, and thus warn all their fellows. Surely, by doing this, they are incurring extra risk? Why shouldn't the individual who first spots the predator, simply creep away and leave the others to it? One possibility – hard to test, but it makes sense – is that by triggering general panic the whistle-blowers *increase* their own chances of survival, because predators are confused by general panic; and besides, if some hapless infant or old-timer is stirred into action it is more likely to be targeted than is the individuals who stay alert and flash the warning. So the conspicuous tail flashers may look very noble, but they could just be setting up their mates as patsies. It's usually not hard to be cynical, if you put your mind to it.

But some examples of apparent self-sacrifice are not so readily explained away – and the most notable of all, which Shakespeare commented upon, is provided by the honey-bee. As T.H. Huxley noted in 1894 (35 years after *Origin of Species*, when expressions like 'struggle for existence' were well established):

> The society formed by the hive bee fulfils the ideal of the communistic aphorism 'to each according to his needs, from each according to his capacity'. Within it, the struggle for existence is strictly limited. Queens, drones, and workers have each their allotted sufficiency of food ... A thoughtful drone (workers and queens would have no leisure for speculation) with a turn for ethical philosophy, must needs profess himself an intuitive moralist of the purest water. He would point out, with perfect justice, that the devotion of the workers to a life of ceaseless toil for a mere subsistence wage, cannot be accounted for by enlightened selfishness, or by any other sort of utilitarian motives.'[10]

Indeed, before the days of neo-Darwinism Mark II such facts were very hard to explain. Even the most buttoned-down of thinkers – and there's been few more buttoned-down than T.H. Huxley – have resorted to speculative moralising in the face of them. In the honey-bee colony, a single queen lays all the eggs. A few idle males, known as drones, are produced when it is convenient to produce them and have nothing to do but impregnate more queens. All the work of the colony – foraging, cleaning, child-care, defence – is carried out by the 30,000 or so workers. Such is their toil that they die from exhaustion within a few weeks – unless, along the way, they lay down their lives in the common cause by stinging some intruder, for the act of stinging rips their bodies apart. Many a sage has seen the honey-bee as a model for us all to follow, embracing all of life's virtues – chaste, hard-working, dedicated to the common cause. Others have seen the hive as an incestuous and generally monstrous stewpot of despotism and slavery. All who have thought about bees have found them puzzling. For what motivates the workers to do as they do?

An obvious answer – or at least partial answer – would be that the workers are simply sterile. The queen, who does all the reproducing, must surely have micky-finned the eggs that give rise to workers, so that they grow up barren. But if that's all there is to it, why don't those putatively sterile workers simply give up – find some nice flower to sup upon, and sunbathe away the time until their hour is done? Why work? Why attack invaders at such personal cost? Why not simply sneak away and live to doze another day? The workers' behaviour becomes all the more mysterious when we discover that in truth they are not sterile at all. They are perfectly capable of producing eggs which, if cared for, turn into drones – which in turn are able to impregnate queens. So workers are biologically able, it seems, to found colonies of their own. Yet they don't. Apparently voluntarily, they sacrifice their own reproductive potential in favour of the seemingly ungrateful queen. If a worker does lay eggs (which occasionally happens) the other workers kill them off. How nasty; or as Lewis Carroll's Alice would have put the matter – 'Curiouser and curiouser!'

But all becomes clear when we take two basic facts into account, and apply the principle of 'kin selection'. This is an old idea in principle but it was given a respectable mathematical basis and brought to the world's attention in the 1950s by Bill Hamilton. The notion seems simple in principle, though surprisingly complicated maths is needed to pin it down.

The first basic fact is that bees have a strange way of determining sex. The females, both the queen and the workers, are 'diploid'. They emerge from fertilised eggs, which means that they inherit one set of genes from one parent, and another complete set from the other parent, just as we and most other animals do. But the males, the drones, are haploid. They develop parthenogenetically from unfertilised eggs. They have only one set of genes and all of them come from the mother. The males are not clones of the mother or of each other because although each male egg is given 50% of the mother's genes, no two eggs are likely to receive exactly the same 50%. So all the males contain exclusively maternal genes, but they almost always have different maternal genes. In fact all of the Hymenoptera – the order of insects that also includes wasps and ants – determine sex in this peculiar way. The females are diploid and the males are haploid.

The second basic fact is that the young honey-bee virgin queen founds a new colony by first mating with twelve to fourteen males – then mixes their sperm thoroughly, and keeps it in store for the rest of her active life. From then on, she can expose her eggs to sperm and so produce diploid female offspring (which mostly turn into workers); or she can lay unfertilised eggs and thus produce a new generation of males.

Now the principle of kin selection chimes in, and so does the maths.

For the principle of kin selection – in line with the general idea of 'the selfish gene' – is that the *real* purpose of reproduction is not to make more honey-bees or human beings or runner-beans or whatever the lineage may be, but to make more genes. We (bees, people, beans) don't make more or our own kind for our own benefit (our offspring potentially are rivals, after all, and the producing and the raising of them tends to wear us down) but because our genes tell us to. If we reproduce sexually, then each of our offspring contains 50% of our genes – which for most creatures, under most circumstances, is the best we can hope for (unless we practise incest, which has dangers of its own). But if we can't have children of our own, then it's a reasonable deal in principle to help to look after your siblings. Siblings share about 50% of your genes and their offspring – your nephews and nieces – each share about 25% of your genes, which is a lot better than nothing. Thus in nature we find that animals of many kinds – including wolves and naked mole-rats and various birds such as ground hornbills and Florida scrub-jays – do indeed help to raise their siblings and/ or their nephews and nieces. Obviously, from the genes' point of view, the closer the genetic relationship, the more worthwhile it is to help with the childcare.

The worker bees are able only to produce unfertilised eggs who grow into males, each of which shares 50% of her genes. In practice that is the best that any of us can hope for (unless we resort to asexual reproduction and get ourselves cloned in the manner of Dolly the sheep, which, even if it was possible, would be extremely risky). So of all the options open to the worker, her preferred option should be to have (haploid) sons of her own.

In a traditional human family of the kind described in eighteenth and nineteenth century novels, it is considered reasonable for an elder daughter to care for her siblings, especially if she herself remains unmarried. Indeed it is reasonable, genetically speaking, because the nieces and nephews that her siblings may give rise to should share about 25% of her genes. But for honey-bees, it's not so. Firstly the workers' sisters are in effect sterile (not literally so; but in practice they don't have viable offspring) so there will be no nephews and nieces. Looking after younger workers even if they are your mother's daughters is a biological dead end. Secondly, because the queen mates with more than one drone, only a minority (about one in 12 or 14) of the workers are full sisters. Most are half sisters. So each worker is liable to have not 50% of genes in common with her fellow workers, but as few as 25%. She will have less than 25% of her genes in common with the offspring of the other workers. But a worker does have at least 25% of her genes in common with the offspring of the queen – for the offspring of the queen are of course her half-sisters (and sometimes are her full sisters). So in practice, the best option for honey-bee bee workers is to help to raise the offspring of their mother, the queen. Most of those offspring will be workers, to be sure, and so are a dead end. But a few will privileged individuals will be brought up to be queens, and so will continue the dynasty – passing on at least a proportion of the genes contained within each worker. It's a pretty raw deal for the workers, whichever way you look at it. Hobson's choice. But Hobson's choice, though not good, is by definition the best there is. In short, the best course for the workers is to do what in fact they do do: help to raise the offspring of their mother. Or that, at least, is the best option for their genes.

Why, though, do workers attack the eggs that are occasionally laid by other workers? Because they have fewer genes in common with the eggs of their half-sisters than they do with the eggs of their mother. So they kill their half-sisters' eggs, and focus on the needs of the queen's eggs. In principle, if they were very clever, the worker bees could form a pact, a trade union, and agree to help each other out: 'I'll help you to raise your eggs, if you'll help me to raise mine'. This is the kind of thing that human beings do – although, as history shows, even humans don't find it easy to maintain such contracts, which is why the ruling minority is able to stay in charge. Alas – although it's lucky for humanity and for all the other beneficiaries of honey-bee industriousness – bees aren't that clever. They go where their genes guide them. They look after their mother's eggs, and destroy those of their half-sisters, and the colony remains intact.

Bumble bees, incidentally, though they have a broadly similar social arrangement, form much smaller colonies than honey-bees, and at the end of the season the new generation of virgin queens fly off alone to start a new colony they do not take an entourage of workers with them, as honey-bee queens do. Why so? Because the bumble-bee queens, unlike the honey-bee queens, are not promiscuous. They mate with only one male at the beginning of the season. So all the bumble bee workers are *full* sisters, with 50% of genes in common. So their genetic loyalty to their sisters is just as great as it is to the queen, so they are far more likely to rebel. That, at least, roughly, is the theory; and it matches the facts, which at least is a good start.

It was Bill Hamilton who first explained the mystery of social insects in terms of kin selection, and it's a wonderful insight. At one stroke it seems to explain one of nature's greatest puzzles. On its own, the example of the honey-bee seems to vindicate the entire thesis that lies behind the notion of 'the selfish gene': that behaviour is guided primarily by the needs of genes, rather than of the individuals who harbour those genes.

What applies especially to honey-bee (because of their peculiar way of determining sex, and their mother's promiscuity) applies in principle to all us of. According to the theory, the great but serendipitous paradox is that creatures are capable of unselfish acts *because* their genes are 'selfish'. Sometimes – often – more replicas of any one gene result if that gene prompts its possessor to help others who also possess that gene, than if that gene encouraged its possessor to attempt reproduction on its own behalf.

Often, though, things don't work out like that. Often, as you might expect, genes are most likely to replicate themselves if they promote the survival of their possessor come what may. Then, as Bob Trivers in particular continues to argue, we may see conflict between individuals even in relationships that seem unequivocally harmonious – not because the individuals bear each other any ill will but because their genes are fighting their own corners.

Indeed, as David Haig has shown, we can see such conflict of interest acted out even in the mammalian womb, between mother and fetus. For example, all of us all the time produce insulin to reduce our blood sugar. 'Hyperglycaemia' is dangerous as all diabetics are aware. But the placenta, which in effect is part of the fetus, produces a hormone of its own called human placental lactogen (hPL) which counters the effect of the mother's insulin. It's in the short-term interests of the fetus, after all, to ensure that the mother's blood sugar is high – for this, for the fetus, is its food supply. The mother is obliged to produce more and more insulin to overcome the hPL but sometimes the hPL dominates – and pregnancy-induced diabetes is the result.

Matt Ridley, in his *Origins of Virtue*, points out that David Haig is not trying to claim 'that all pregnancies are tugs of bitter war between enemies'. Nonetheless, says Ridley:

... as well as the shared genetic interest between them,

> there are also some divergent ambitions. The mother's selflessness conceals the fact that her genes act as if motivated entirely by selfishness, whether being nice to the fetus or fighting it. Even within the inner sanctum of love and mutual aid – the womb itself – we have found ruthless assertion of self-interest.[11]

More subtle (nature is capable of endless subtlety) is the phenomenon of 'meiotic drive' – now known in a wide range of the creatures that are the laboratory favourites, including mice, fruit flies, maize, and the fungus *Neurospora*. Presumably (or at least very possibly) it is universal. Meiosis is the process by which the two sets of chromosomes of the germ cell mix themselves up and then separate to form gametes, each containing just one set of chromosomes. The germ cells of the male just divide neatly, so that each germ cell gives rise to two equal spermatozoa. But in the female, the germ cells divide unevenly. Only one of the resulting offspring cells goes on to become an egg. The other, called the polar body, is discarded.

Normally in meiosis each gene in each germ cell has an equal chance of winding up in either of the two daughter cells. In the male line all is well, because both daughters of the male germ cell go on to become spermatozoa, capable of fathering the next generation. But in the female germ cell, half the genes are unlucky. Half of them get shuffled in to the polar body, and discarded.

In the form of 'meiotic drive' that interests us here a gene in the female germ line operates in such a way that it increases its chances of getting into an egg, rather than into the polar body; so it is much more likely to be passed on to the next generation. This perhaps would not matter so much if the gene that cheated in this way was of a kind that does no harm. But some genes that practise meiotic drive are positively harmful. Some cause sterility in the males of the next generation. Some kill the embryos. In populations of wild mice (bearing in mind that a single haystack may be big enough to hold a viable population of wild mice) the spread of such genes has been known to lead to extinction.

In short, deleterious genes that practise meiotic drive

demonstrate gene selfishness in two ways. First they compete with their fellow genes to increase their own chances of being passed on. Then they damage the offspring that they finish up in. This is selfishness writ large. Combine this hard-line version of neo-Darwinism with the Lucretian-style materialism that we identified in the opening chapter, and we have what might be called 'ultra-Darwinism'. Indeed, Richard Dawkins has referred to himself as an 'ultra-Darwinist'.

Ultra-Darwinism seems wonderfully logical. Clearly, there are facts to support it, and supportive facts by definition count as evidence. Evidence based on observation plus sound logic equals science, does it not? So we might not like the conclusion – the strictly material universe driven by all-out-competition between its components is not a cozy picture – but if that's where the science leads us, then that must be the truth, and we just have to get on with it.

But does it? Is it? Do we?

Are genes really 'selfish'?

All of the above is good science. The facts, painfully extracted over decades, do indeed seem to be as they are described here. The behaviour of all creatures, including us, is to at least some extent influenced by our genes. Although this has been controversial of late it is self-evidently so, at least at a simple level; for if I had the genes of a dog, say, I wouldn't be writing this, and if you had the genes of a giraffe you wouldn't be reading it. 'Altruism' can be explained as a mechanism that enhances the replication of particular genes; the explanation fits the facts and the maths is surely sound. There is indeed tension between parents and offspring, including mother and fetus. Sometimes genes do seem to do open battle with their fellow genes – and sometimes then they may harm their possessors, all apparently in the interests of their short-term replication. So at least in a broad, arm-waving way it seems reasonable to suggest that life is indeed driven by genes that are bent solely on their own replication, come what may. Here, we are given to understand, is Darwinian natural selection – paraphrased as 'survival of the fittest' – played out at the molecular level.

But it is important to see where the science ends and the rhetoric begins. We are indeed influenced by our genes but is it helpful – is it *right?* – to suggest that life is simply a hierarchy, ruled by a boss called DNA – to say as Richard Dawkins famously did:

> We are survival machines – robust vehicles blindly programmed to preserve the selfish molecules known as genes ... They are in you and me; they created us, body and mind; and their preservation is the ultimate rationale for our existence.[12]

There is tension in the womb to be sure, and between parents and children (which indeed is a prime theme of literature, as portrayed not least in King Lear) but is David Haig right to suggest (at least as paraphrased by Matt Ridley) that:

> In all sorts of ways ... the fetus and its slave, the placenta, act more like subtle internal parasites than like friends, trying to assert their interests over those of the mother.[13]

Is it fair to assert, as George Williams did, that:

> As a general rule, a modern biologist seeing an animal doing something to benefit another assumes either that it is being manipulated by the other individual or that it is being subtly selfish.[14]

At least, if 'a modern biologist' does take such a view is he or she justified in doing so?

The answer to all of these questions is 'No'. We could say that the emphasis on competition that lies behind the metaphor of the 'selfish' gene is Darwinian but we need not assume (great though Darwin unquestionably was) that it is therefore correct.

The descriptive 'selfish' is a shameless piece of anthropomorphism – and although I will argue later that anthropomorphism has an important place in biology, this is not where it belongs. In practice the rhetoric springs not from science (science *qua* science doesn't do rhetoric) but from an assortment of pre-conceptions of a philosophical, political, sociological, and poetical nature (of the kind that influenced Darwin himself). There's the Enlightenment belief, first of all, that it is good to think mechanistically – that an explanation based on molecules must trump any other kind of explanation. There's the in-built conviction, etched deep in western culture even without Darwin, that life is one long struggle for personal supremacy – summarised in Tennyson's all-too resonant phrase, 'Nature red in tooth and claw'. There's also a belief, never made explicit but always present in western academe, that intellectuals know best: that ideas must trump intuition; and ideas with maths behind them must be best of all.

Biologists are said to suffer from 'physics envy', and some do. They want to deal in irreducible fundamentals, summarised in maths. But the biologists who think like this are envying the physics of two hundred years ago, which dealt in certainties and straightforward cause and effect in a perfectly predictable universe – all of which now seems naïve. Worse, biologists who envy physics have misconstrued the nature of their own subject. For while physics deals with fundamentals of matter and energy, biology looks at the world as those fundamentals interact, which they do in an infinity of ways. In truth biology has more in common with literature than with physics. In great novels or indeed in soap operas a host of sub-plots interweave – and so it is in life. The competitiveness of life is only one theme among many. It just happens to be the one that Darwin lighted upon – and is easy to express mathematically.

All this is illustrated in the next few chapters: that although competition is an inescapable theme of life the essence of life is cooperation. Life is not a punch-up. It is a dialogue – and a constructive dialogue at that. If it were not, there would be no life at all.

3. The Dialogue Of Life

The modern neo-Darwinians – 'ultra-Darwinians' – have left us with a seriously bleak view of life. They present it as a to-the-death struggle, one long punch-up. But it's not a heroic conflict between gods, or of wild but noble beasts fighting for their right to live and for their children. It's a behind-the-scenes, no-holds-barred, merciless yet dispassionate battle between mindless, faceless scraps of an organic acid known as DNA. We human beings, and tigers and beetles and oak trees and toadstools are mere puppets, all of us driven by our selfish genes. For good measure (although this has nothing directly to do with the case) Richard Dawkins, who coined the expression 'selfish gene', tells us that the universe as a whole exhibits 'nothing but blind pitiless indifference. So we are all alone in this universe and surrounded by enemies. This is supposed to be 'reality' – and we must face up to this reality for the alternative is to live atavistically in a dreamland, populated by fairies, or simply to bury our heads in the sand and wile away our unexamined lives. It is not only stupid to reject or even to question this ultimately pessimistic view of life, we're told. It is dishonest and therefore positively ignoble. The ultra-Darwinians have taken to calling themselves 'the brights', meaning that those who disagree must be the dims. The brights have rooted their ideas in science, and science cannot be wrong. Game, set, and match.

But when you look at life with true dispassion, Darwin's 'struggle for existence' and Spencer's 'survival of the fittest'; when you just look at life as it *is*, it doesn't really look like one long struggle at all (as Darwin himself commented). Natural selection is surely a fact of life but it need not imply competition. Competition certainly need not imply conflict. The science itself, in which the ultra-Darwinian thesis is rooted, has been misrepresented. Neither does it make sense – no sense at all – to say that genes are 'selfish'. That is mere anthropomorphism, not of the creative and heuristic kind (which, I sill argue later, is essential) but of the kind we have long been warned against. We cannot hope to understand our fellow creatures – seals and elephants and crows and other such beasts – without being anthropomorphic, at least in a disciplined way. But anthropomorphism applied for rhetorical purposes to non-sentient entities, like lengths of DNA, is merely sensationalist. Sensationalism too may well have a place in human discourse – but not in the context of what claims to be science.

In short, the whole thesis of the selfish gene, sitting right at the heart of what passes as modern biology and indeed of modern rationalism, is in truth an uneasy amalgam of dubious scientific interpretation and rhetoric. Scientists and philosophers who aspire to be our intellectual leaders really should know where science ends and rhetoric begins. Just to anticipate an argument I want to pursue later, the wholesale rejection by 'the brights' of anything that smacks of metaphysics – any suggestion that there may be more to life and the universe than they can contain within their dogmas and their algorithms – is nothing more than their personal point of view, and is in no way privileged. These particular clothes of these particular emperors are distinctly threadbare.

But we will come to that. First let us look at the basic biology.

Is life really so competitive – and does competition mean conflict?

Natural selection is one of the great insights of all time. It has spilled over from biology and now is properly seen as a general principle that applies to all aspects of life. Admittedly, some biologists in recent decades – and, indeed, virtually from the time that *Origin* was published – have doubted whether natural selection really is as important as Darwin suggested. Many have pointed out that some of the events that have done most to shape the course of life on Earth had nothing very directly to do with the particular mechanism of natural selection at all. After all, about 65 million years ago, at the end of the geological period known as the Cretaceous, a giant asteroid hit the Earth somewhere around present-day Mexico and threw up a cloud of debris dense enough to obliterate the rays of the sun – and so precipitated a change of climate that apparently put paid to the dinosaurs. The dinosaurs had been the dominant mega-fauna on land for the previous 120 million years while the mammals (which in truth are a very ancient group, at least as old as the dinosaurs) had lived as hole-in-corner creatures very much in their shadow. If the asteroid hadn't polished off the terrestrial dinosaurs at the close of the Cretaceous, we mammals might still be living in their shadow, and creatures like you and me very probably would never have evolved at all. Where stands natural selection in the face of what in effect is simply chance?

Yet there are various kinds of answer to this. First, Darwin himself stressed, from the outset, that natural selection almost certainly is not the only cause of evolutionary change – so we needn't get too worked up when we find other forces at work. Then again we could argue that the survival of the mammals and the demise of the terrestrial dinosaurs in the face of the asteroid is in truth a fine example of natural selection in action. Very big land vertebrates disappeared in the post-asteroid 'winter', probably through lack of food, while small ones – the mammals – were able to survive in micro-niches. We may note, too, that some dinosaurs did survive the asteroid – in the form of birds. Birds, genealogically speaking, *are* dinosaurs: they are loosely related to *Velociraptor*, made famous by Stephen Spielberg in *Jurassic Park*, and are in the same general group as *Tyrannosaurus rex*. Physiologically and ecologically, though, as small-bodied, warm-blooded creatures, the birds of the late Cretaceous were far more similar to the late Cretaceous mammals – and they survived the catastrophe happily enough.

Finally, I once had the opportunity to discuss this whole issue with John Maynard Smith, one of the four biologists whose work was summarised by Richard Dawkins in *The Selfish Gene*. Maynard Smith simply pointed out that however profoundly life may be shaped by asteroids and rising mountains and dwindling seas and all the rest, nothing else of a natural kind can explain the unquestionable fact that living creatures as a whole are so beautifully adapted to their surroundings, and so beautifully adapted to the presence of other creatures of their own and of other species. Hamlet observed that 'there is a divinity that shapes our ends, rough-hew them how we will'; and we might in naturalistic vein substitute 'natural selection' for 'divinity' and speak of 'life's exigencies' rather than 'we'. (But I do not intend that the idea of natural selection should *replace* divinity. There is room for both, as Darwin himself constantly intimated and many a scientist and cleric since has been more than happy to acknowledge. I will discuss this later).

Anyway, let us accept, as most post-Darwinian biologists have, that natural selection plays a significant and indeed a decisive part in shaping the course of evolution. Then the question is – how does it really work?

As we have seen, Darwin began with the idea that he derived from Malthus – that all creatures have the potential to out-breed their resources and so there must be a constant struggle for food and space, which means they must 'struggle for life'. The word 'struggle' implies conflict – and indeed Darwin also spoke of the 'fight' for life. But is life really such a struggle? And does natural selection really imply that struggle is inevitable? And does 'struggle really mean 'fight'?

To begin with, we can observe that natural selection happens even when there is no great Malthusian struggle at all. The first living systems to appear on Earth had the place to themselves – yet they very obviously evolved. (I say 'living systems' rather than 'organisms' because, as we will see, the first life on Earth probably took the form of what has been called 'the primordial (or primeval) slime'. It took a long time and a lot of refinements to evolve any kind of entity that could qualify as a bona fide organism). Similarly, the first vertebrates to evolve hard skeletons in a sea full of softies, or the first to evolve jaws in a sea full of suckers and scrapers, or the first to venture on to land, found themselves with a clear field. Yet in such circumstances they evolved very rapidly, in each case into a huge variety of forms. Still of course we could say that the each new class of creature in its virgin environment was competing up to a point – though not, initially, with other creatures of a comparable kind. As Darwin himself said, the principal obstacle in any situation is not other creatures, but the environment itself. The priority for any creature (or any living system) in all circumstances is first and foremost to *survive*. Life in general is a hard trick to pull. The first task for all of us, human beings and toadstools and staphylococci alike, is to get through the next 24 hours, or indeed the next second.

To put the matter more mathematically, we could say that the prime task for a living system – whether it is some very ancient piece of slime, or an amoeba, or an elephant, or a single gene – is to survive from time T_0 (T-zero – the starting point) to T_x (where T_x is any time after T_0 – from a millisecond to eternity).

If we made a list of all the living systems that were present at time T_0, and then came back and visited them again at time T_x, we would find that some of them had fallen by the wayside. Bland and obvious though that statement is, it is, in the end, all we are fully justified in saying about natural selection. Some entities survive better than others. We can't, without guessing, say *why* some survive better (as Darwin himself pointed out). We can merely observe that some do better than others; as the adage has it, 'nothing succeeds like success'. Neither need we envisage that the different entities are engaged in any kind of struggle with each other. They might indeed have absolutely no contact with each other. Still we would get the same result: some survive better than others.

Let's call the survivors *A*s, and the ones that are not so lucky, *B*s. Now let us imagine (which is true, after all) that living entities have a tendency to replicate. Clearly, nothing can replicate unless it first survives. So at time T_x we wouldn't simply find that the *A*s had survived and that the *B*s had disappeared. We would find that the *A*s had multiplied. A mixed population of *A*s and *B*s would have been replaced by an exclusive monoculture of *A*s. So it looks to suspicious and over-analytical minds as if the *A*s have pushed the *B*s aside. It looks after all as if the *B*s had perished *because* the *A*s had multiplied and gobbled up all the resources. But this would merely be an inference. The evidence is circumstantial. We have no right to assume, without further, direct evidence, that the *B*s had actually been actively done down. We are justified only in observing what is obvious – that some entities, the *A*s, have survived and replicated, while others, the *B*s, have perished and so (obviously enough) have not replicated. In fact, we infer that the *A*s have shoved the *B*s aside only because that inference matches our preconceptions. Our preconception in this case has been honed by Malthus, Tennyson, and Darwin. If one group has flourished while another has perished we take it to be self-evident that the former must have ousted the latter. But actually, there is no a priori reason at all to assume that this is so.

Two examples from real life will make the point. When I first studied biology formally in the 1950s no-one knew about the asteroid of 65 million years ago. But they did know – it had been known since the nineteenth century – that the terrestrial dinosaurs had disappeared suddenly from the fossil record at the end of the Cretaceous, and that sizable mammals had appeared soon afterwards. So it was argued (I remember writing essays about it) that the small and scrawny mammals of the late Cretaceous had somehow ousted the mighty dinosaurs. Although the mammals of the time were pea-brained it was assumed that *because* they were mammals they must be brighter than any dinosaur, because dinosaurs were crudely classed as 'reptiles' and it was taken to be self-evident that mammals must be smarter (because, after all, we ourselves are mammals). One theory had it that while the parent dinosaurs were gazing vacantly into space or munching the topmost leaves of ancient conifers, their heads almost literally in the clouds, the neat-footed mammals nipped in and nicked their eggs.

Now it is clear (and it should have been clear then) that the later dinosaurs were comparatively large-brained, even if they didn't quite have the strategic powers of three-star generals as depicted in *Jurassic Park*. Their fossil footprints and other kinds of evidence show that some at least had a complex social life. It is also known now (although it wasn't known in the mid twentieth century) that the later dinosaurs (and very probably a lot of the earlier ones) were very good parents. Entire fossil nests complete with eggs and brooding mothers have been found. The late Cretaceous mammals must have been content to work around the mighty and increasingly intellectual dinosaurs, just as their ancestors had been doing for aeons.

In short, the story which said that there was conflict between the two great phylogenetic lineages, the dinosaurs and the mammals, and that the mammals won – with the inference that they did so because they were more like us – was pure invention, and almost certainly pure nonsense. It was entirely based on a kind of 'Darwinian' preconception, liberally laced with anthropocentric imperialist conceit.

The second example dates from a mere two to three million years ago: the great Pliocene-Pleistocene interchange of beasts between North and South America. Now the two great continents are uneasily joined via the isthmus of Panama, but for a very long time it wasn't so. South America began as part of the great southern continent of Gondwana, attached to the land masses that now form Antarctica, Australia, New Zealand, and Africa. North America was part of the great northern landmass of Laurasia, together with Eurasia and Greenland. South America began to break loose from the rest of Gondwana about 180 million years ago and at least by 30 million years ago it was an island, slowly drifting north. During that time it evolved an extraordinary array of mammals which included a great many marsupials – though generally quite distinct from those of Australia. Some of the South American marsupials are still with us but most, including one that for all the world resembled a sabre-tooth cat, have disappeared. South America also had its share of placentals, including armadillos and anteaters which belong to the placental group known as the edentates, which used to live in other continents as well but now are confined to South America. Somehow too – it's not clear how – the island of South America acquired its own suite of rodents which include including guinea-pigs and capybaras; plus monkeys (the New World monkeys including the spider and woolly monkeys); and small cats (ocelots and so on). South America in its island form also had a range of large herbivores, some of which resembled horses while others were very like camels and yet others were like elephants – although they clearly belonged to quite different lineages which had and have no counterpart anywhere else. (This is a fine example of 'convergent evolution' of which more later).

But between two to three million years ago, at the end of the Pliocene epoch and leading into the Pleistocene, the drifting island of South America finally made contact with North America and so gave rise to the map we have today, or very nearly. Then, unsurprisingly, there was a great exchange of animals, including mammals, between the two continents.

But the exchange was very uneven. Big cats, including pumas and jaguars, poured down from the north. Bears flowed down too – ancestors of the spectacled bear which still lives in the Andes. Elephants in the form of mammoths and mastodonts came in as well. Horses came in, including three-toed kinds – though all are now extinct. (The modern horses of South America were introduced by Europeans). Camels believe it or not are a North American invention and although there are none left now in the north (apart from a few modern types imported from North Africa and Asia) South America still has its quartet of 'camelids' – the llama, guanaco, alpaca, and vicuna.

In contrast, only a few of the southerners flowed successfully north, including the marsupial opossums and the placental armadillos, now common in North America, and the giant ground sloths, which flourished for a time but were finally seen off some time after 13,000 years ago, apparently by early human invaders. South American monkeys, ocelots, several guinea-pig like rodents (including pacas and agoutis), and tree sloths have made it as far as Central America, but no further. Furthermore, as the northerners flowed in to South America, what was left of the native southerners, including all the big indigenous herbivores, died out. All this is clear from the fossil record.

The standard 'Darwinian' explanation, as with the dinosaur-mammal story, is one of conflict and victory. The dim southerners missed out to the smart northerners – who must be smart because they are still with us, while most of the southerners aren't. We can imagine (it isn't difficult) true horses and elephants from the north gobbling the vegetation before the native southern look-alikes could get to it. The big native herbivores had clearly survived the depredations of the big marsupial sabre-tooths but succumbed nonetheless to the quicker wits and agility of jaguars and pumas. It all fits the fact perfectly. There must have been a punch-up, and the northerners won. There can be no other explanation.

But there is another explanation, suggested some decades ago by the South African zoologist Elisabeth Vrba (now based at Yale). She simply pointed out that as the Pliocene gave way to the Pleistocene, which is when the two Americas collided, the world as a whole was growing rapidly cooler. Much of the north had been covered in lush, semi-tropical forest. Now it became sparser, grassier, and less lush. To find the kind of forest they were adapted to, the animals of the north had to go south. If the southern continent had not joined up with the north when it did, then many of the northerners may have perished – shoved southwards into the sea. With Panama in place, they could carry on migrating down towards the Equator, which runs right through the centre of Brazil. In contrast, the native animals of South America had always lived on the Equator, or thereabouts. They thrived in forest at its most tropical. As the world cooled, their own, native, hyper-tropical forest retreated. It was replaced by forest that was less dense, less lush – precisely the kind that the Northern species were adapted to.

In short: we do not see a straight fight between northerners and southerners. We see one group of animals (the northerners) tracking a form of vegetation that they are adapted to, as it moves south. Meanwhile another group (the southerners) are increasingly deprived of the kind of vegetation that they have evolved to deal with. So the northerners, as the climate cooled, were able to stay in their comfort zone just by moving house: while the southerners found that their favoured habitat was fast disappearing.

If there was a contest, it was a most uneven one: conditions changed to favour the invaders. More to the point, we have no reason to suppose there was any kind of contest at all. Even if we concede that there was competition between the northerners and the southerners, we certainly need not infer that there was any kind of direct conflict between the two. We simply have to envisage that both groups fed as best they could on whatever plants were growing – and the plants that now were growing in the cooling climate were more like those of the erstwhile North than of the erstwhile South. Unsurprisingly, the animals that were better adapted to the new fare that was on offer, did better than the ones that were less well adapted. I am reminded of the advice I once heard (on television) from Michael Johnson, the great American sprinter. If you are running the 200 metres, he said, don't look at your opponents. Don't spend your time thinking that you are running against *them*. Just focus on the line, and on your own technique. I reckon that this is what most living creatures do, most of the time. They don't think about out-doing each other. Metaphorically speaking they simply run their own race, and hope for the best.

It all hinges, I reckon, on the definition of the word 'competition'. Darwin took it to be self-evident that life is a struggle and then in conflated the general 'struggle for life' with the conflicts that take place between different creatures. The rivalry between gazelles that seek only to get on with their lives and the cheetahs that seek to hunt them down is indeed a competition, and indeed is conflict. The clash of mountain billy-goats as they seek to lay claim to the females is also competition, and indeed is conflict. For all the participants in these various forms of fracas, conflict mode is essential, including the rush of adrenalin and sometimes of testosterone that feed the fires of aggression.

But it seems to me positively perverse to suggest that a lizard warming itself in the sun to speed up its physiology, or me sitting here typing, both of us basically seeking only to do the things that life requires, are involved in any kind of what can properly be called competition. Certainly in neither case is there any sense of conflict. We *could* say (as Darwin indeed did) that the act of staying alive and minding one's own business is itself a form of competition, since one is constantly 'struggling' (in a manner of speaking) to keep death at bay. But this, surely, is to twist the meaning of the word 'competition'; and as many a philosopher and some theologians have demonstrated this past few thousand years, we can prove anything we like if we twist the meaning of words as we go along. I suggest indeed that getting on with one's own life, and competing with other creatures, are two quite different things. It is a simple mistake to conflate the two. Life can be seen to be innately competitive only if we suggest that doing nothing and wiling away the time is itself a competitive act. The notion that life is inescapably competitive becomes positively absurd if we assume, as is so often assumed (and is implicit in the word 'struggle') that competition necessarily implies conflict. It is downright misleading to imply that the acts of survival necessarily involve selfishness – that selfishness is a necessary and inevitable survival tactic. In fact, as Darwin himself pointed out, it is usually impossible to say why or how one particular lineage of creatures survives rather than another. All we can properly say is that the survivors survive: that 'Nothing succeeds like success'. There is no reason at all to assume that the creatures or genes that do best – at least in all but the shortest term – are in any worthwhile sense of the word 'selfish'.

This brings us to what I see as a huge serendipity: that in practice, the best survival tactic of all is to cooperate. Life, in reality, is not a fight. It is a dialogue. Dialogue may involve competition to be sure. But the essence of dialogue is cooperation. We see this clearly when we look at the origins of life itself, and indeed at the nature of life.

The nature of life

Sooner or later in all respectable courses of biology the question must be raised – so what exactly is this thing we are studying? What *is* life? How does a dog differ from a stone – or from a statue of a dog? Lists are made: living things grow; they reproduce; they move and feel (at least sometimes).

Then the conditional clauses creep in. Stalactites grow but we don't say that they are living. Crystals grow too – and they break up, and then each bit may go on growing, so we could say that they reproduce. But do they live? The sea moves – but if ever we say that it's alive it's only by way of metaphor.

So then the lists grow more subtle. Living things have a very complicated structure – lots of little bits, and lots of different molecules, all acting together to make a whole, each dependent on all the others. This is one meaning of the word 'organic'. Stones, by contrast, (and statues made out of stone) are essentially crystalline: one kind of molecule is repeated over and over again, with no complex interactions between them. A crystal is a static thing. In absolute contrast, the cytoplasm of a living cell is constantly interacting. If it's not visibly on the move then it's probably dead. Life is not a *thing,* in fact. It is, as the polymathic Jonathan Miller once put the matter, a *performance*; more like a flame; a perpetual to-ing and fro-ing. Indeed, a living structure is constantly re-creating itself, even when, like the dusty aspidistra on the boarding house window-sill, it seems to have given up. To maintain all this complexity and activity the living thing (unlike the dullard stone) is constantly using energy. The never-ending use of energy and the constant self re-creation – in fact the sum total of all the busy things that go on in a living system – is what's meant by *metabolism.*

In recent decades – in the age of neo-Darwinism – the discussion has taken a different turn. Some have said that we should define life in evolutionary terms. Any system that seems to evolve by natural selection and hence to become better adapted to its surroundings could (or indeed should) be considered to be alive – or so some scientists have argued. Some crystals can take many different forms as they grow and multiply, and some forms survive better in some circumstances than others. So why shouldn't we say that the crystals are adapting to their surroundings by natural selection? And if they are doing that, why shouldn't we say they are alive? We can design computer programmes that can take many different forms depending on conditions, and invade other programmes and take them over – just like the viruses we find in nature. So why (some have asked) don't we say that they are a form of life?

Chemists, on the other hand, at least for the past two hundred years, have sought, as indeed they might, to define life in terms of chemistry. The basic components of flesh are made of carbon, oxygen, hydrogen and nitrogen, with significant inputs from phosphorus, sulphur, sodium, potassium, calcium, magnesium, iron – and indeed, when you add them up, from about half the periodic table. But the element that is always present in all living things, and the key player, is carbon. Indeed, when chemists say 'organic' they don't mean complex interacting systems where everything depends on everything else, as artists and designers (and most of us) generally mean. They simply mean 'contains carbon'. So that seems peremptorily to rule out computer programmes (which these days conventionally depend on silicon).

In ultra-sophisticated vein, neo-Darwinians Mark II have commonly narrowed down the essential chemistry to that of DNA. As we've seen: DNA is seen to be the boss, the sovereign, the point of the whole operation. All the rest, including all the general metabolism of the cell, is presented to us as the below-stairs staff. For DNA, it is observed, is the molecule above all that replicates in a very orderly fashion, and is the main target of natural selection. If evolution is to happen at all, then it has to happen primarily at the level of DNA; and if DNA changes, the rest changes too. Life originated, we are sometimes told these days (I have heard professional scientists say it) when DNA began. DNA, after all, we are told, was the 'first self-replicating molecule'; and is the one on which all else depends.

The business of building proteins according to the code

encapsulated in DNA is complex, and in practice it is mediated by a whole family of assistant molecules known as RNAs. RNAs are very similar chemically to DNA but they are simpler: they have only one strand, instead of two strands that are intertwined to form the double helix. They are also more versatile: some at least are able to act as catalysts, as enzymes do. For these and other reasons it is now widely accepted that RNAs have evolved earlier than DNAs, and generally acted as jacks-of-all-trades before DNA came on the scene, and was able to provide the codes for a whole army of specialised proteins in the form of enzymes. But the general idea persists: life began when some key molecule – DNA or RNA – with the power to multiply itself and to boss the rest around, came on board.

Inevitably, too, underlying this neo-Darwinian view of things, the ever-present leitmotiv, is the idea of competition. From the outset, we are given to understand, the replicating molecules were battling with each other for resources and *lebensraum.* In short, in the ultra-modern ultra-Darwinian view, life began as a punch-up, and DNA emerged in very short order as the leader of the pack; and, in the four billion years since, each molecule of DNA has in the end, despite appearances, been engaged in a to-the-death struggle with all other molecules of DNA. For all our angst and conceits, we and our fellow creatures are just the vehicles of DNA, the buses and trucks and ships and aeroplanes, the deluded cannon-fodder, or indeed the by-standers (choose your own metaphor).

No scientist that I know has ever stated the modern view of life in quite the terms presented above. But the above, nonetheless, is a fair paraphrase of a very great deal of what I have heard and read about the origins and nature of life over the past four decades and roughly summarises what entire generations of biologists have been brought up to believe. It is the essence of what is now perceived as neo-Darwinism, and has sometimes been called ultra-Darwinism.

If this ultra-Darwinian account were true it would be bad news indeed. Life and indeed the universe are perceived as one big punch-up and we ourselves, and all other living creatures, emerge as by-products. Fortunately, though, the ultra-Darwinian view is mostly junk; flawed through and through, to its core. We need to re-think from basics: about what life is really; about how evolution really works; and how, in practice, life could have begun.

What life is really, and how it became

Just to re-cap: the two big things that living systems do and non-living things generally do not is to metabolise, and to reproduce. (I say 'generally' only because some crystals and computer viruses and so on may reproduce after a fashion). Ultra-Darwinians in recent decades have focused on the second of these – the ability and the apparent urge of living things to reproduce. They have focused in this way firstly because they have tended to *define* life as whatever can evolve; and if evolution is to happen – 'descent with modification' as Darwin put it – then replication is a prerequisite. In particular, as we have seen, the modern neo-Darwinians have focused on the replication of DNA. Indeed, many have argued that life began when DNA came on the scene and started to replicate.

But the ultra-Darwinians are putting the cart before the horse. More fundamental by far, and indeed the essence of life, is metabolism: all the things that go on in the cell besides the replication of DNA.

So what is the essence of metabolism? At first glance, it's nothing very fancy. The essential feature is nothing more nor less than the chemical cycle. A chemical cycle is like a chain reaction but with the end joined up to the beginning. In a chain reaction, *A* makes *B* makes *C* makes *D* and so on. So you start with *A* and you finish up with something else. There are many examples – one of which is the destruction of the ozone layer in the upper atmosphere triggered by molecules of chlorofluorocarbons (CFCs) that float up from discarded refrigerators (among other things). But in a cycle, *A* makes *B* makes *C* makes *D*, and so on and so on – but whatever is next-to-last in the chain, say *X*, makes more *A*. So the cycle goes round and round and round and finishes up with *A* again.

A cycle that just did this would be of very little interest. But the kind of cycles that really make a difference to the world incorporate more material as they go around – and finish up making more A than they began with. They also tend to spin off various bits and pieces of other things as they proceed. A prime example in the living cell is the Krebs cycle (also known as the citric acid cycle or the Szent-Györgyi-Krebs cycle). It is found in all organisms that practise aerobic respiration (meaning that they respire with the help of oxygen). In detail the Krebs cycle is ludicrously complicated (as metabolic pathways and life in general tend to be) but the net result of it is to burn sugar (let's say glucose) with the aid of oxygen, in an extremely controlled way; so that the energy released from the sugar doesn't just disappear in a flash of heat and with an alluring whiff of caramel but instead is incorporated (after many a further transformation) into a molecule of ATP (adenosine triphosphate) from which it can be gently released when required by removing one of the Ps (that is, reducing adenosine triphosphate into adenosine disphosphate, *aka* ADP). In practice, though, the sugars fed into the Krebs cycle are first broken down to form pyruvic acid (pyruvate), which then goes through about nine transformations before finishing up as pyruvate again. Each transformation is catalysed by its own specialised enzyme. Complex though it is, the Krebs cycle clearly evolved a very long time ago.

The job of the Krebs cycle is to generate energy. Clearly, though, it could not continue to operate unless it also *used* some energy. The Krebs cycle generates more energy than it uses but most of life's cycles and the metabolism in general use more energy than they generate. The Krebs cycle (and others like it) are specialised mechanisms that power the rest of the metabolism.

So here we have it. The essence of life. The essence of life is not DNA. It is the chemical cycle of a kind that is able to keep cycling forever and ever by snaffling energy from the outside world.

We can indeed compare it to a flame – but it's a flame that has acquired the art of self-refuelling.

It has sometimes been remarked that life seems to defy the laws of entropy, which says that energy always runs down, and that orderly systems (which require energy to maintain their order) tend to become disorderly. Yet life seems to *gain* energy as it progresses, and becomes *more* orderly. But of course life does not break the laws of entropy. The dogma of physics says that nothing can break the laws of entropy. Living systems acquire the energy they need by robbing their environment. Organisms that photosynthesise (plants, seaweeds, and many bacteria) take energy from the sun. Animals take energy from sugars that have been made by other organisms, and initially are made by organisms (such as plants) that practise photosynthesis. But long before there were animals or plants or photosynthesising bacteria there were microbes of many kinds that obtained energy by reducing or oxidising compounds of sulphur or of iron and other things too. There are many thousands of different kinds of reaction in nature that release energy – and different organisms that have appeared over the past four billion years have made use of a whole range of them. Some still do: in places where there is sulphur and iron, we find bacteria and archaeans exploiting their chemistry, in various ways. Photosynthesis has become so pre-eminent only because light is more ubiquitous than any one chemical deposit; and once photosynthesis has happened, the creatures that practise it become an easy source of energy for creatures of many different kinds, including bacteria and fungi and animals, that have evolved to consume them.

So where should we look for the origins of life? Obviously life did *not* begin with DNA. The essence is the cycle, and we can envisage cycles of many kinds – acquiring energy from their surroundings and so able to continue in principle forever and ever – without DNA. A cycle that cycled potentially in perpetuity would be expressing the essential feature of life: not a particular thing but a kind of a process – a robbing and endless re-cycling of energy through a series of chemical entities. Such a system needs no DNA to keep it going. Indeed it has long been obvious that DNA is a highly evolved molecule – and one which emphatically is *not* self-replicating. It needs a whole entourage of proteins (and other bits of nucleic acid) to enable it to replicate. In other words DNA could not have evolved at all *except* in an environment that was already very much alive. So DNA is not the begetter of life. It is a product, a scion, of life. RNA is simpler than DNA and presumably appeared on the scene before DNA. But RNA too is highly complex and highly evolved, and needs help from its friends, and could not have come into being except within a system that was already alive. DNA, in short, and RNA too, are Johnnies-come-lately to the living world.

Note, too, crucially, the *nature* of early life. It certainly could not have begun with one particular molecule which then created an entourage which it bossed about. The initial cycles must have arisen as cooperatives of different molecules that had quite separate origins. The kinds of molecules that are surely relevant would have included amino acids, which nowadays are the raw materials from which proteins are made; and nucleotides, which are the building blocks (as the cliché has it) of DNA and the various forms of RNA. Some of them presumably arose spontaneously on Earth – bearing in mind that the Earth, at the time they first appeared, about four billion years ago – was a very different place from now. Notably its atmosphere contained no oxygen, but reeked instead (had there been anything around to appreciate the reeking) of ammonia and methane and (so some have speculated) with hydrogen cyanide and other highly potent gases that most of today's creatures (though not all) would find intolerable. Because there was no free oxygen, too, there could be no ozone layer – so the still-young Earth would have been bombarded with cosmic radiation (which the ozone largely excludes). The moon was much closer and huge in the sky and the tides must have been horrendous, exposing vast stretches of land and then dousing them again, while volcanoes raged and hot springs were commonplace. Altogether it was an alchemist's dream, a stew-pot of the most potent chemistry. But then again, some of the essential ingredients of life may have been born in outer space, and brought to earth in meteors. There is evidence for this.

So life arose in the midst of turbulence – yet it did not itself arise as a turbulent Darwinian struggle but as a collaboration between quite different kinds of molecule that began to interact and then, after many a trial and error (there were plenty of molecules and plenty of time to play with) began to form cycles. As we will see, life is still in essence collaborative. If it were not so, then life could not exist at all.

Clearly, to shift from primordial slime – a lot of carbon-based cycles milling about – to the kind of systems we now conventionally acknowledge to be living, required a huge amount of evolution. Popular accounts (and even textbooks) sometimes tell us that bacteria are the most primitive form of life but in truth they are highly evolved and sophisticated organisms. We might say in rhetorical vein that by the time bacteria or something like them came on the scene then evolution was all over bar the shouting. The oldest known bacteria (or bacteria-like organisms) are about 3.5 billion years old – but that was not the beginning. The Earth is 4.6 billion years old and a huge amount must have happened in that first billion years.

What *exactly* happened in that first billion years is almost anyone's guess, however. All we can reasonably do is to list the kind of changes that must have happened, in what we feel might be a plausible order, and then wave our arms in the way people do when they feel they ought to say something but are not sure of their ground.

So first, we might reasonably guess, cycles evolved of the kind already described. Then over time natural selection would ensure that some of those cycles became more efficient – not because there was a punch-up between rival cycles, but because the ones that worked better had a better chance of survival. In practice – not necessarily, but in practice – increase in efficiency is likely to imply increase in complexity. We can imagine that some of the molecules taking part in the cycles would change over time (in a chemically active soup there is plenty to change them) and that some of the altered types would work more efficiently than the original types, and hang around – commonly interacting and so cooperating with what was already there.

Growth is not an essential prerequisite of survival, as we have seen, but systems that grow are more likely to survive than those that do not. Bigger systems with more molecules on board also have more chance of growing more complex and efficient, at least in parts. Soon we can envisage that natural selection would favour separation: the bits of the system that worked best would work even better if they became cut off from the bits that did not work so well. Once we have separated systems natural selection favours replication – because several or many copies of a system that works well have a greater chance of surviving. When complex systems replicate mistakes creep in. It is hard to replicate a complex system accurately. So this – at last! – is where RNA and then DNA come into their own. In all living systems, proteins are the key players. They form the enzymes which catalyse most of the reactions of the metabolism. Proteins, as we saw in the previous chapter, are in dialogue with nucleic acids – RNA and DNA. The structure of the nucleic acids reflects that of the proteins, and vice versa. The nucleic acids (in contrast to the proteins) have a simple structure and are easy to replicate accurately. Thus they reflect the structure of the proteins not by imitating the proteins, but in the form of an easily replicated code. As molecular biologist Sydney Brenner put the matter recently in *Nature:*

> ... every organism contains an internal description of itself. The concept of the gene as a symbolic representation of the organism – a code script – is a fundamental feature of the living world and must form the kernel of biological theory.[15]

In short, RNA and then DNA evolved – within living systems that had themselves been evolving for many million years – as a data store. They help the living system to remember how to make appropriate proteins. DNA is not, in short, and never was, the boss. It was, and is, the librarian.

Of course the above is arm-waving. It would take tens of thousands of words to put substantial flesh on the bones and send everyone to sleep (I have tried it). It would also remain inescapably speculative. But *in essence*, the above is a plausible account of the kinds of things that must have happened. We start with fairly inchoate metabolism in an unprepossessing slime and we finish up with coherent systems that grow and replicate in an orderly fashion and are kept on track by the ultra-sophisticated data-storing molecule known as DNA. Those coherent systems can properly be called 'cells'. The whole process is *not* primarily competitive. It is cooperative. The essence of the relationship between DNA and the rest of the system is *not* one of master and servants, but of dialogue. DNA itself cannot replicate without help from a great many other molecules, among which proteins are key players (and of course, DNA is a key player in creating those proteins in the first place). Emphatically not is DNA 'self-replicating'. To achieve anything at all, DNA needs help from its friends. That last comment is of course both anthropomorphic and rhetorical. Yes indeed: but no more so than 'selfish gene'. 'DNA and chums' reflects the reality of life at least as accurately as 'selfish gene' although the connotations are clearly very different.

So it is that in *The Selfish Gene* Richard Dawkins described the life of genes as follows:

> Now they swarm in huge colonies, safe inside gigantic lumbering robots, sealed off from the outside world, communicating with it by tortuous indirect routes, manipulating it by remote control. They are in you and me; they created us, body and mind; and their preservation is the ultimate rationale for our existence.[16]

Powerful stuff. But although it concerns a subject best dealt with by science it is not actually a scientific statement. It is rhetoric: a literary, eloquent but highly metaphorical, statement of an opinion. But whereas we like to think that the scientific facts of the case are solid (or at least, we try to make them as solid as possible) the rhetoric we attach to those facts is eminently flexible. No-one has demonstrated this more neatly than the Oxford cardiologist and theoretical biologist Denis Noble in *The Music of Life*. Thus he re-presents Dawkins's account of genes as follows:

> Now they are trapped in huge colonies, locked inside highly intelligent beings, moulded by the outside world, communicating with it by complex processes, through which, blindly, as if by magic, function emerges. They are in you and me: we are the system that allows their code to be read; and their preservation is totally dependent on the joy we experience in reproducing ourselves. We are the ultimate rationale for their existence.[17]

Dawkins and Noble are talking about the same thing – the relationship between genes and the cells and organisms of which they are a part. But the stories they tell are as different as those of plaintiff and defendant in a court of law. The question that lawyers are obliged to ask – which story is *true*? – is not really appropriate. Dispassionately, we could argue that both stories are true up to a point. Both express aspects of the truth.

But conventional modern biologists are wont these days to favour the Dawkins-style account. They seem to prefer the idea that life is basically conflict. Thus if we say – 'Life seems to me to be highly cooperative!' – the modern biologist is wont to say, 'Yes, but, it is *basically* competitive! Genes, cells, and whole organisms collaborate with each other only as a matter of convenience, because there is strength in numbers and that makes it easier to bash the rest!'

But why is the competition perceived to be more 'basic' than the collaboration? If we look as far as we are able at life's origins, we see that it could not have begun *except* as a collaboration. If we explore the relationship between genes and genes, and the genes (DNA) and the rest of the cell, we find it is one of dialogue: a to-and-fro of information. Nothing can survive without everything else. Every living thing that aspires to exist at all must master the arts of survival – but the best survival tactic by far, and indeed the *sine qua non,* is to cooperate. Of course dialogue involves competition up to a point. But why is competition seen to be more fundamental? In truth (if we look at the facts dispassionately) we see that while competition is a fact of life – in most circumstances, a more or less inescapable fact – collaboration, cooperation, is life's *essence.* I think this should be written in letters six-feet high. Life is *essentially* cooperative. If it was not, it could not work.

I suggest that modern conventional biologists stress life's competitiveness not because this is sound philosophy and science but for reasons that are historical and sociological. The moderns have been brought up to be neo-Darwinists. The whole 'Darwinian' thesis in truth as we have seen is has many component threads – Thomas Hobbes, Enlightenment rationality, Malthus, Tennyson, and Herbert Spencer as well as Darwin himself. It has become the principal *leitmotif* of western politics and economics. It has the force of dogma. Life's fundamental competitiveness is taken to be self-evident. Life's obvious collaborativeness is taken merely to be an epiphenomenon, or a Machiavellian device; a temporary tactic to help us to compete more effectively. The dogma is not to be questioned. Yet it is eminently questionable. In the end, the stress on competition rather than on collaboration is a matter of rhetoric. Some scientists are very good rhetoricians. But it important to know where science ends and rhetoric begins and not to mistake the two.

I will end this chapter with a brief account of some of the most stunning insights of the twentieth century – insights that show life's essential collaborativeness beyond all reasonable cavil.

How creatures like us became

In the 1970s an American biologist, Karl Woese, showed that the creatures which mid-twentieth century biologists lumped together as 'bacteria' in fact belong to two fundamentally different groups. One is the bacteria. The other is the archaea. They look much the same under the light microscope and in nature their functions overlap but, said Woese, in key aspects of their basic biochemistry they are fundamentally different. The colloquial term 'microbe' is still useful, however – and I will use it now and again in the following to mean bacteria and archaeans collectively.

I said above that by the time evolution had produced creatures as intricate as bacteria, then it was all over bar the shouting. After all, although modern microbes (including both bacteria and archaeans) tend to be specialists, each occupying its own particular niche, between them they exploit just about every conceivable environment and can nourish themselves and respire in an extraordinary number of ways. Hot springs, fierce ores, hyper-saline seas – bring them on: some microbe or other can cope. Among other things, one group of bacteria – known as cyanobacteria; once wrongly called 'blue-green algae' – invented photosynthesis. But then again, although each bacterium and archaean is a specialist they can and do as a matter of course swap genes between them, and/or entire sequences of genes, and so acquire each others' skills. In fact they practise various forms of what might properly be called sex. They have DNA of an advanced kind, which enables them to replicate themselves in an orderly and generally accurate fashion (and often at extraordinary speed. Some of them in favourable circumstances have a generation time of minutes). So although bacteria and archaeans are often said to be 'primitive' and 'simple' in truth they are miracles of biological efficiency and miniaturisation.

But although microbes are wonderful they have limitations. In particular, although they have many skills between them, and can borrow skills from each other, each particular kind of bacterium or archaean tends to be highly specialised. Individually, they are not versatile. In addition, although some bacteria do form colonies in which the whole is greater than the sum of the parts, they are condemned by the nature of their anatomy to be small. Yet there are advantages both in multi-tasking and in being big (not the least of which is that big things can eat little things. Of course life is competitive up to a point).

The kind of body cells that we have – and mushrooms and oak trees and 'protozoa' – are far more complex, and more versatile than those of microbes. Most obviously, we keep our DNA neatly bundled within a discrete nucleus. Bacteria and archaeans don't have a nucleus. Their DNA is packaged in various ways throughout the body of the cell. Our cells also contain specialist organelles: a variety of bodies within the cytoplasm, each of which carries out some specialist task. Notably, the cells of animals and plants (in fact, all creatures that respire with the aid of oxygen) contain mitochondria, which deal with aerobic respiration. Plant cells in addition have chloroplasts, containing chlorophyll, the green pigment that captures light energy from the Sun and drives photosynthesis. Creatures with such complex, multi-functional cells are called 'eukaryotes'. Many eukaryotes (the ones loosely called 'protozoa') operate as single-celled organisms. But in the ones that we see all around us the cells remain attached to each other as they replicate and so form multi-cellular organisms – which of course can be vast: as big as giant sequoias or blue whales.

So how did eukaryotic organisms, creatures like us and giant sequoias, come about? This is where twentieth century biology turned previous thinking on its head. In 1905 a Russian biologist Konstantin Mereschkowski first proposed that eukaryotic cells had arisen as coalescences between different kinds of microbe (actually he wasn't quite the first to suggest this but he seems to have put the idea on the map). This suggestion didn't seem to attract the attention it surely deserved until Lynn Margulis in the United States took up the notion in 1960s – and she continued to develop it up until her death in 2011. Then in the 1970s Karl Woese pointed out that the DNA and the biochemistry of eukaryotic cells has affinities both with bacteria and with archaeans.

Putting all the ideas together it now seems that eukaryotic cells first arose as coalescences of archaeans and bacteria; and then the new combined entities took a whole series of other bacteria on board which evolved into the organelles and other structures. Most obviously, the chloroplasts of plants evolved from cyanobacteria which took up residence as 'commensals' (lodgers) and then stayed on and became integrated; and the mitochondria, shared by almost all eukaryotes including both plants and animals, evolved from resident purple bacteria. Biologists continue to argue about the details. But the general idea ('endosymbiosis') is now widely accepted. In fact it can reasonably be seen as the orthodoxy.

So the individual body cells of which you and I (and oak trees and mushrooms and all the rest) are composed are absolute exemplars of collaboration. The collaboration is compounded and then compounded again as our body cells (and those of oak trees and the rest) multiplied, stayed together, and then specialised to form liver or brain or leaves or roots or whatever is appropriate; combining to form organisms that are far, far greater in their capabilities than any one isolated cell could possibly be. In short, we and all the creatures that are big enough to see with the naked eye, are masterpieces of collaboration.

Of course we can still argue if we like, as Richard Dawkins does, that the marvellous collaboration that is an oak tree or a human being is just an extrapolation of a selfish molecule striving to reproduce. But such a view is not, in itself, science. It is rhetoric. It is a point of view, a 'take', eloquently and somewhat chillingly expressed. We might suggest, though, in the light of the now acknowledged facts, that this point of view is somewhat perverse. The real question is not, 'Is it true?' but 'Why should anyone argue such a thing?' The answer, I suggest, is not a matter of intellectual purity. The answer lies with sociology, and the history of the western world, and in particular with the thread of thought that runs from the Enlightenment (beginning with Thomas Hobbes) through Malthus and Tennyson to Darwin and Spencer, and onwards into a materialist, logical positivist, hard-nosed, buttoned-down species of science.

Overall, we can envisage life in general and the universe as a whole as a balance between competitiveness and cooperativeness. Thus, at least in whimsical fashion, we can see a parallel with the worldview of the Zoroastrians or the Manicheans – or indeed with the particular Christian tradition presented by John Milton in *Paradise Lost*: constant tension between evil (competition) and good (cooperation). We should stress, however, that as in *Paradise Lost*, cooperativeness (good) must prevail – or life and the universe could not exist at all. Perhaps indeed this isn't so whimsical at all. Perhaps those ancient theologians got it right all along. Clearly, though, the ultra-Darwinian view of the world has serious limitations – and so too as we will discuss in Chapter 10, does the hard-nosed, buttoned-down version of science that has given rise to it.

For the next two chapters I want to show how different life looks when we begin with the idea that it is fundamentally co-operative.

4. Nice, Clever Animals And Nice, Clever People

In the end – although predation and wars and general viciousness are facts of life – life in essence is collaborative. If it were not so, life would not be possible at all. Taken all in all, and in most circumstances, collaboration is the best survival tactic – and so we should expect Darwinian natural selection to favour collaborativeness. In the case of animals, which interact largely through physical contact, with all the hazards that this entails, we would expect natural selection to favour sociality: the ability to get along with others of one's own kind at a *personal* level (and sometimes with other creatures of different kinds). We would also expect that as evolution proceeds, the nature and the scope of the social relationships would become more and more intricate and more subtle. On the whole, this is precisely what we do find.

No animal is an island

Greta Garbo famously wanted to be alone and a great many animals including Sumatran rhinoceroses, tigers, and giant pandas seem to prefer solitude most of the time. But even the most stand-offish need help from others of their own kind, at least from time to time. Some animals can produce offspring parthenogenetically (meaning they practise virgin birth, which they do in a variety of ways) but most practise sex at some stage of their lives – and sex above all requires cooperation. Flowering plants and conifers and suchlike contrive to practise sex without direct physical contact – delivering their male gametes packaged as pollen via the wind or by courtesy of bees or humming birds or lemurs or whatever. Some animals, including a great many marine invertebrates, can also achieve fertilisation at long range, by squirting their gametes into the plankton – but even they need to coordinate their efforts, so that sperm and eggs are around at the same time.

But in most animals that reproduce sexually the partners need to make physical contact, and this means they need to find mates, and this requires some degree of sociality. So even those that look anti-social, like giant pandas, tigers, and Sumatran rhinos, are really far more social than they look. All those beasts spread themselves out primarily for ecological reasons. Sumatran rhinos and giant pandas live on low-grade vegetation, and they need a lot of space if they are to find enough, and tigers hunt alone and would get in each others' way if they all hunted over the same ground. But all of them surely know by clues of scents and sounds who their neighbours are, and where they live, so they are not socially unaware; and tigers in zoos have sometimes proved to be remarkably sociable, once the pressure to find food is removed.

Of course, there are degrees of sociality. Woodlice cluster under damp stones simply because they all like cool damp places. We need not assume strong social bonds between them. Yet close contact of a fortuitous kind can be beneficial – such that baby birds and mammals, born into the same nest, may benefit from each other's body warmth (such as it is). When any kind of behaviour brings any kind of advantage, natural selection is likely to favour it, and build upon it. So we find in thousands of species, in thousands of different ways, mere proximity has evolved into true interaction, with benefit to all parties.

First though we should distinguish between animals whose social behaviour seems primarily to be hard-wired, such as ants, bees, and termites; and animals that clearly rely heavily on learning, and have some degree of behavioural flexibility, like starlings and zebras and human beings. There's overlap between the two, of course. Some biologists have argued that bees and ants are far more aware of what's going on than most of us would suppose; and some mammals – even the kinds that we suppose are very clever, including ourselves – are more stereotyped than we might care to suppose, which suggests some degree of hard-wiring. But the distinction holds nonetheless; and in this account I'm speaking only of creatures such as starlings and zebras and human beings who, one feels, are not just slaves to their genes, and whose nervous circuitry allows them some flexibility.

The benefits of sociality extend across the board, into every aspect of life. So we find musk-ox, those ancient shaggy half-cattle half-mega-sheep from the north, warding off wolves by gathering in a circle, their formidable horns facing outwards, with the young ones sheltering in the centre. So too we find small birds of many kinds, often in mixed flocks with several species, combining to mob birds of prey. In these examples the defence is physical, head on. But sometimes the defence comes from increased vigilance – many pairs of eyes are better than one. Then whoever spots the danger first may simply flee and the rest follow – but often the first to run away does so conspicuously, so that the others can see the warning. So it is that rabbits on the move flash their white tails, and many drab and camouflaged birds from redwings to golden plovers burst into colour when they take flight – confusing predators and alerting their flock-mates. Sometimes the vigilance seems highly organised; and so it is that mammals such as meerkats and birds such as the Arabian Babblers post sentinels, who seem to run great personal risk as they stand on high places to watch out for aerial attack, and issue the warning when danger threatens. (Amotz Zahavi's studies of Arabian Babblers are a joy).

Many animals coordinate their breeding and thus produce a huge number of offspring all at once – so many that the predators are over-faced, and many of the babies that otherwise would be sitting ducks, pull through. So wildebeest and other meaty beasts on the plains of Africa give birth all together, producing more babies than the lions, leopards, and hyaenas are able to slaughter in the time available. Some animals won't breed at all unless there are many others of their kind all around them, to help the process along. If zoos want to breed flamingoes they have keep a fair-sized flock of them (and when they have only a few they have sometimes tried to fool the birds by setting up mirrors).

Many animals forage or hunt together – with demonstrable increase in efficiency. Vultures and condors whirl in the sky in groups – and they spot more corpses that way than they would on their own. They also look out for other scavengers feeding, and muscle in or wait their turn depending on who got there first. But even here, the feeding isn't merely competitive. Vultures can't generally feed on the corpses of big animals until some lion or hyaena has first torn a hole through the hide; and the marabou storks of Africa, committed scavengers yet without the hooked beaks of vultures and the great jaws of lions and hyaenas, cannot sensibly get a look in until the corpse is well dissected. Lions, wolves, and wild dogs demonstrably have a higher success rate than solitary hunters like tigers and leopards. Of course, when hunters who work in teams do catch something, they have to share it. On the other hand, they are better able to protect their kill against robbers – hyaenas, jackals, or whatever – than they could if they were on their own. As every *aficionado* of TV wildlife movies knows, cheetahs, heroic, high-speed but solitary hunters that they are, constantly lose their hard-won booty to lions and hyaenas.

All animals that practise sex have to form some kind of relationship with a mate, however brief. But some need extra help from third parties. Giant pandas do indeed lead solitary lives but when a female is in season (an event that occurs only for a few days a year) the males move in from miles around. Female giant pandas are not easily aroused, but the attention excites her: she is far more likely to be receptive when a whole mob of males is vying for her favours. Probably this is one reason why it's so difficult to breed pandas in zoos. Zoos rarely have more than a pair and without male rivalry for stimulation, the females tend to be frigid. Some animals actually cooperate in mating. Male long-tailed manakins – small, usually bright-coloured birds from the American tropics – perform their mating dance as a double-act: one of them is the prospective mate, while the other serves as his assistant. But the assistant is really an apprentice. After a few seasons as a helper he finds an apprentice of his own and becomes the boss, the one who reproduces.

In many birds – much more so than in mammals – child-care is shared between both parents. Sometimes, too, in both mammals and birds, the parents are helped out by older siblings or by other relatives – and occasionally by individuals who are not related at all. In the ground hornbills of Africa the babies in the latest brood are cared for in part by siblings from an earlier brood, who later go on to have children of their own (having already learnt some of the wiles of parenthood). In some other cases the helpers rarely or never get to do their own breeding. The breeding strategy of naked mole-rats – hairless rodents that live colonially in burrows in the drier parts of Africa – resembles that of termites. Only the queen breeds. She is much larger than the rest, who all serve as workers, helping to protect her young. Eventually, when she dies, one of the workers puts on a spurt of growth and takes over, while the rest carry on working. The sudden growth is mediated by hormones, and production of hormones in general is greatly influenced by state of mind. So long as the worker mole-rats are overseen by the heavyweight queen, the appropriate hormones are not produced.

In all these cases – and we could easily cite a hundred more – it's easy to see how the different participants benefit from collaboration even when their role is subordinate, and they seem to be exploited by their social superiors. The usual explanation, which usually works perfectly well, is kin selection: the individuals who seem to serve only as helpers are related to the bosses. This is where the selfish gene hypothesis really does seem to apply. Young ground hornbills help their parents to rear more offspring. This means they help to rear their own siblings – who, of course, share a proportion of their genes; so the genes in the helper in truth are helping to care for copies of their own kind. As we saw in Chapter 2, kin selection seems even to explain the extreme sociality of honey-bees which (we may assume) rely more or less absolutely on their genes to tell them what to do.

Yet as we will see shortly, kin selection does not explain everything. Clearly in nature (and certainly among domestic animals and animals in zoos) there are real friendships between unrelated animals, sometimes even between animals of different species. Crucial to all this is the very obvious relationship between sociality and intelligence.

Social animals are the brightest

'Intelligence', 'awareness', 'consciousness' – these are elusive concepts perceived and defined in many different ways. Yet they are clearly related, and overlap. Creatures deemed intelligent are able to learn and also to behave flexibly: to solve puzzles; at least to some extent to weigh up the situation and decide on one course of action rather than another; and able to communicate a variety of ideas. 'Conscious' implies some degree of self-consciousness – the conscious creature, human or otherwise, knows that he or she is actually doing something. Mere complexity of behaviour does not always signify intelligence or consciousness. Ant behaviour is remarkably complex but ants are not usually thought to be intelligent or truly to appreciate what they are doing and why.

But true intelligence comes at a price. It requires a complex nervous system coordinated by a brain. In general, the bigger the brain (relative to body size) the more intelligent a creature is likely to be. But nervous systems in general and brains in particular are high-maintenance. They require an awful lot of energy to keep them going. Yet they have to be cost-effective. There is no point in having a great brain, using several times more calories weight for weight than normal flesh, unless that brain offers a pay-off that is at least commensurate with its upkeep.

So what is the pay-off? Well, you might reasonably suppose that it is always useful to be more intelligent. But is it? Suppose some mutant fish appeared with the literary powers of Jane Austen. What good would it do? Even if the fish was able to find some coral-free piece of rock on which to inscribe the maritime version of *Pride and Prejudice* ('It is a truth universally acknowledged that a flounder ...') who would read it? Who would reward it for its efforts? Besides, while the literary super-fish was scratching away, a bigger fish would eat it. No: on the whole, fish are better off being fish – alert to predators, attractive to potential mates (or otherwise able to have their way), and either equipped with enormous jaws or very well camouflaged or able to move very fast. None of these essential piscine attributes requires great brain power. In fact, we can assume that for most human beings too, through most of history, the genius of Jane Austen would have been surplus to requirements and often a positive burden.

There have been many explanations of animal – and human – intelligence and different ones may apply in different circumstances. The very special intelligence of human beings (and the commensurately enormous brains) seem to need at least two kinds of explanation. First, the traditional idea, as first mooted formally in the nineteenth century, surely applies: brains and hands co-evolved, each spurring on the other. Thus the traditional (Darwin-inspired) account of human evolution suggests that we first acquired gripping hands (with opposable thumbs) and flexible arms because our ancestors, for at least the previous fifty million years, lived in the trees. Then, some time after three million years ago, the global climate started to cool and East Africa grew even cooler than it otherwise would as tectonic pressure forced the land mass upwards. As the world cooled it grew drier and so the forest retreated, and so the apes that became our ancestors spent longer and longer on the ground. They began to walk upright – bipedal walking is very efficient for those who can do it, and it gives a better view – and this freed their hands. Their forest ancestors were already making tools – as indeed do modern chimps, our nearest living relatives – and with hands now free, our ancestors were able to make even better tools than their arboreal ancestors had done.

In turn, the liberated and already dexterous hands of our apish ancestors provided the pay-off for bigger brains. With tools of wood, bone, horn, and stone, our almost-human ancestors were able to crack nuts they couldn't otherwise have cracked, kill animals they couldn't otherwise have killed, and then butcher them, using the meat far more economically than they otherwise could. So there was established a positive feedback loop – the kind of interaction that produces big changes very rapidly. Better tools meant a more reliable and more nutritious food supply with which to nourish bigger brains which in turn provided the wit to make even better tools. The fossil record shows that the brains of our ancestors increased from chimp-sized to human-sized – at least a three-fold increase – between about three million and somewhat less than one million years ago. It's an astonishing change. But unless the increase in brain size had been immediately rewarded, with more and better food, this could not have happened.

Yet human beings are not the only animals that deserve to be called intelligent. The more that biologists observe our fellow creatures, the cleverer they find them to be. Monkeys, apes, pigs, elephants, whales, squirrels, dogs and wolves, hyaenas, crows, parrots, starlings and even some fish seem to have extraordinary mental powers. But only a few of them use tools, and very few indeed actually *make* tools, so the kind of mechanism that plausibly prompted our own brains to grow so spectacularly cannot apply to most other animals. There must be a more general spur to intelligence that applies to animals of many different kinds. Human intelligence, we may suppose, was driven by this general spur in addition to the special effect of hand-brain co-evolution. So what is it that in general encouraged and encourages increase in brain-size, and hence (broadly speaking) in intelligence?

According to Robin Dunbar, now at Oxford, and Nick Humphrey, now at the London School of Economics and Harvard, the answer is – sociality. The general advantages of sociality are obvious and legion but to get the full benefits and avoid internal conflict, there has to be some structuring. In a herd, it often helps if the most experienced animals take the lead – and so we find many a matriarchy, among elephants and deer and sheep. In any group of animals, the youngsters need to be taken care of but they also need to know their place or they will mess up the hunts and expose the group to predators and fail to learn the necessary skills of life. Social hierarchies develop – what's commonly known as the 'pecking order'. This order commonly depends largely on who is related to whom. Among chimps, the offspring of high-ranking females have far higher status than those of low-ranking females, and a far greater chance of survival. In all kinds of contexts it is alarmingly easy to draw parallels between chimps and humans.

Sometimes it is hard to see the advantages of such formal social arrangements, at least for the runts and the tail-end-Charlies in the group who get kicked around. Yet the advantages generally outweigh the injustices. Low ranking chimps may constantly be harassed and beaten up but at least, in a group, they have some chance of surviving and reproducing. On their own they would have no chance at all. In this perhaps we see the root reason why so many human societies cling to the status quo even when the government and the other ruling groups are obviously incompetent. The alternative to group living, however unpleasant the group may be, is death; and governments, however self-deceptively, invariably claim to increase the coherence of the group.

Social coherence requires every individual to know every other individual – where each one stands in the hierarchy, who should be feared, and so on. Social animals are remarkably good at this. Zebras look all of a piece to outsiders – yet each member of a group knows the others by the individual pattern of stripes on their ample rumps. Recent studies show that sheep can recognise the individual faces of at least eighty other sheep, which is more than a great many human employers seem to manage. Robin Dunbar has shown that the more intelligent the animal, the bigger the social group it can cope with – and from a survival point of view, big groups commonly have the edge over smaller groups. The optimum size for a human working community, he says, is 150. Above that, and the individuals start to lose their sense of who's who.[18]

In short: sociality in animals brings huge advantages; and sociality requires intelligence (unless the animal is hard-wired to form a particular kind of social group); and it also encourages the further evolution of intelligence and of whatever is meant by consciousness.

But are non-human animals really intelligent? Haven't we been told, these past few hundred years, that animal intelligence, and particularly animal consciousness, is an illusion? Isn't it merely anthropomorphic to suppose otherwise?

Are animals really clever?

At school and university in the 1950s and 60s we, students of biology and of experimental psychology, were not allowed to say that animals are 'clever'. That was 'anthropomorphic' and anthropomorphism was the sin of sins. It means: crediting non-human entities, including other animals, with attributes including states of mind which are generally presumed to be exclusive to humans. Engraved on my memory is a letter I read at that time in a learned journal from a schoolteacher who warned her fellow pedagogues not to allow their hapless pupils to say that cuckoos lay eggs. What cuckoos do, she shrieked, is 'exhibit egg-depositing behaviour'. Young people these past few decades have largely abandoned science in favour of English with Spanish and media studies. They do this, we are commonly told, because science is too difficult. That doesn't seem to me to be the only explanation.

The aversion to anthropomorphism has multifarious and sometimes ancient roots. Part of it clearly reflects a general desire to protect the idea of human specialness. Thus Genesis tells us that 'Man' was made in the image of God, with the strong implication that the rest were not. The King James translation tells us that 'Man' was given 'dominion' over the rest. Many a cleric – and many a scientist – rejected Darwin's idea that humans and apes share a common ancestry largely on such grounds. Many (including some scientists) accepted the idea of evolution in general, but nursed the idea nonetheless that human beings *were* specially created. In some strands of Christian theology (strands that in truth derive only loosely from Biblical teaching) close relationships with other animals were seen to be somewhat blasphemous (such that pet cats, labeled 'familiars', were sometimes taken as evidence of their owner's witchery).

With impeccable Gallic logic René Descartes assured us in the seventeenth century that animals cannot possibly think because thought depends on verbal language and animals don't use words; and if they can't think they can't have proper emotions either because emotions require us not only to experience sensations but to know that we are experiencing sensations, and such knowledge requires thought, which animals don't have. David Hume in the eighteenth century suggested that animals can feel the crude emotions of fear, hunger, and lust but lack subtlety. Depression, elation, pride, jealousy, grief, and all the rest are beyond them. Anyone who has kept a dog or had close dealings with other bright creatures knows the lie of this. Dogs brood, Hamlet-like. Baby elephants are marvellously humorous. But philosophers, from the depths of their armchairs, have presumed to know better.

To be fair, though, there were firmer reasons in the twentieth century for avoiding anthropomorphism. There was a broad and general desire from Sigmund Freud to William James to Ivan Pavlov to turn psychology into a bona fide science. The alternative, it was felt, was simply to be literary: to describe and ultimately to explain the to-ings and fro-ings of animals in much the same way as Shakespeare or Chekhov or Tolstoy described and analysed the foibles of human beings. But what could science do, that these literary giants do not? Well, as always, that science must deal only with phenomena that can be observed reliably, and measured, and subjected to statistical analysis. *Anna Karenina* is a wonderful insight into human feelings and behaviour but there is nothing in it that can meaningfully be measured.

In fact, it is extremely hard to measure the thing that most people think that psychology is supposed to deal with – which is state of mind. At least, such measurement is possible up to a point in humans, largely through questionnaires. But with animals, which cannot answer questions posed by human beings with clip boards, that route is far harder to pursue. What can be measured in animals, though – and in humans too of course – is what they actually *do*. Their behaviour. Hence, roughly at the beginning of the twentieth century the school of psychology known as 'behaviourism' was born. It was soon to be bolstered by the fashionable school of philosophy known as logical positivism, which said that any question that could not definitively be answered was 'meaningless' and should not be asked at all. Questions about animal consciousness and the subtleties of animal emotions were pointless too, since they could not be directly explored. By the 1950s and '60s behavourism was at its height. I studied experimental psychology at university without ever using the words 'thought' or 'emotion' at all.

The two leading lights of the behaviourism in the early twentieth century were first J.B. Watson who developed the idea that complex behaviour can be explained as a series of reflexes; and then B.F. Skinner who argued that behaviour is 'shaped' by rewarding behaviour that works, and punishing what does not. Watson went on to develop advertising techniques which clearly work (though this doesn't mean that they are good, but it does suggest that the underlying theory is not all bad) and Skinner showed that it is more effective to shape behaviour by rewarding desirable behaviour than by punishing bad behaviour – a finding that may have helped to encourage more humane treatment of prisoners.

Then, alongside, in the early to mid twentieth century the science of 'ethology' took off – the formal, quantified study of animal behaviour, not in mazes and puzzle-boxes, but in the field. Pioneers were the Dutchman Niko Tinbergen and the Austrian Konrad Lorenz – great friends (though temporarily separated by World War II) who shared a Nobel Prize in the early 1970s. Both borrowed concepts from the behaviourists, notably that of the reflex, to throw light on the behaviour of wild or semi-wild creatures from sticklebacks to herring gulls (Tinbergen) to geese and parrots and dogs (Lorenz).

But from the mid-1960s onwards, it became more and more obvious that although behaviourism had done a useful job –

ridding psychology of its literary cobwebs – it simply was not adequate. However modified, it could not, in practice, explain or in some cases even begin to explain all that animals get up to when they are given half a chance to express themselves: when they are not simply thrust into a maze with cheese at one end and an electric shock at the other.

Notably, Jane Goodall in the 1960s began to observe wild chimpanzees in Tanzania and showed that their behaviour was in many ways far too complex to explain away as a string of reflexes. In fact it was impossible to get a handle on what they did and how they fared without using concepts that up until then had been taboo. Individual chimps clearly had individual personalities. They were not simply stereotypes of their species. Furthermore, the personality of key individuals in the group – the alpha males and the matriarchs – clearly influenced the overall fate of the group. Thus, individual groups had individual histories heavily influenced by personal foibles – just as is true of human societies. More and more studies of other kinds on many different species – wolves, monkeys, hyaenas, elephants – reached similar conclusions: that behaviourist explanations, based on the idea that animals are simply complicated clockwork toys as Descartes had envisaged three hundred years earlier, would not do. In the 1980s Hal Marcovitz wrote *Behavioural Enrichment the Zoo,* showing how the lives of animals in zoos could be hugely improved by giving them puzzles to solve, and allowing them to make choices. Unashamedly, he wrote of animals feeling confident or depressed and of a whole range of creatures including ostriches, archetypal bird-brains, taking 'pride' in solving problems.

In the 1980s, primatologist Herb Terrace of Columbia University New York declared that the old idea that animals can't think because they don't speak, was nonsense. The problem, he said, is to show how they think *even though* they don't have verbal language. At about the same time Pat Bateson declared that anthropomorphism when applied with restraint was 'heuristic': an essential tool for understanding. In other words, we should not begin by assuming that animals are simply clockwork toys, or computers, until proved otherwise. We could just as well assume, for starters, that at least in their emotional responses other animals are like human beings, until and unless proved otherwise.

The only real caveat – obvious, but worth stating – is that anthropomorphism is good and useful only when it is part of a serious endeavour to understand the animal itself, often with the long-term aim of improving its lot. Obviously it is foolish and potentially cruel to assume that other animals are like us in every detail. It is not kind or sensible to dye a miniature poodle pink and fit it with a diamante collar, just because its owner likes to think that it enjoys being cute. Let the same poodle off the lead and it may chase rats and sheep as vigorously as any wolf (although with far less guile).

But aren't we getting carried away? Are animals really that subtle?

Cleverness in practice

Cleverness does not merely imply complexity. It implies flexibility and adaptability; the ability to respond to any one situation in a variety of possible ways, and to introduce novelty. It may not pay to be too clever, as shown by the fable of the literary fish; and in the modern world, animals not known for their wits, such as tortoises, may do very well. All in all, though, we can see that cleverness is likely to pay – that a big brain would pay its way. So we would expect that natural selection would favour cleverness; and so it often has.

Cleverness – true intelligence – at the very least requires complex neural wiring and is surely a hard act to pull, evolutionarily. Such a quality, we might assume, could surely have evolved only once. Yet high intelligence (and consciousness) have clearly evolved time and time again, independently, in widely separate lineages.

Is it really so? After all, the two obvious stars, mammals and birds, though they have now diverged so far, did share a common ancestor. Perhaps that common ancestor was itself intelligent, and simply passed on its gifts to its descendants. But the most recent common ancestor of birds and mammals lived more than three hundred million years ago and – so the fossil record tells us – it was a shambling reptile with a very small brain. So its descendants, the birds and mammals, must each acquired their intelligence much later, long after they had gone their separate ways. Thus, although intelligence may seem a very hard act for evolution to pull, it clearly happened more than once, independently.

Within each of the two great dynasties, too, of birds and mammals, intelligence clearly evolved independently more than once – in fact many times. For the very earliest mammals, dating from more than two hundred million years ago (the mammal lineage is at least as old as the dinosaur lineage) still had very small brains. So too did *Archaeopteryx*, commonly supposed to be the first bona fide bird, which lived around 140 million years ago. In fact it's clear that the different living mammals that we know are clever – elephants, whales, dogs, primates, squirrels, and so on – must all have evolved their intelligence independently of all the others. So too did the different groups of recognisably clever birds.

Neither does it end there. Among vertebrates, some fish are proving to be brighter than has been assumed – and star among them, oddly enough, is the manta-ray (and rays are distant relatives of sharks, commonly billed as 'mindless' killers, though of course they are no such thing). Among invertebrates, insects are not truly clever, for all their complexity, but some molluscs definitely are. The molluscs that most of us know best – snails and slugs, mussels and oysters – have limited intellects or else have more or less abandoned brains altogether. But the cephalopods are molluscs too and they are remarkable. In the mid twentieth century the great British zoologist J.Z. Young compared the intelligence of octopuses to that of dogs. Squids have a complex language, based on flashing colours, by which they convey a wide range of emotions to other squids. Some say that cuttlefish, generally considered the most primitive of the trio, are the brightest of them all.

This simple and undeniable fact – that significant, bona fide intelligence has arisen independently many times among animals of widely different type – raises several highly intriguing questions of science, philosophy, and indeed of metaphysics. Are there *qualitative* differences between the intelligence of different groups? Do intelligent birds, for instance, look at the world differently from mammals, and process information in different ways? What manner of thing *is* intelligence? Is it simply something that arises when nervous systems (or computers?) grow sufficiently complex? Or is it a general quality of the universe which all creatures may evolve to partake of, just as all (or a great many) have evolved to respond to the universal presence of light? This seems to me a question of supreme (metaphysical) importance (of which more later).

Meanwhile, it seems clear as Darwin suggested that there is no vast and unbridgeable gap between us and other animals. Real friendships grow up between people and parrots, or dogs or elephants; and between dogs and parrots or even cats and budgies. Creation is not divided into us and them, as Genesis is sometimes taken to imply. Truly other animals are our fellow creatures. This is affirmed by our commonality of genes – we are all literally related – but also by our shared intelligence. We all drink at the same pool.

In practice, though, do animals do anything to suggest that they are more than the clockwork toys that Descartes took them to be?

Wordsmiths and puzzle-solvers

Leading non-human intellectuals, it is broadly agreed, are the great apes: chimpanzees, bonobos (formerly known as 'pygmy chimps'), gorillas, and orang-utans. Commonly we acknowledge a hierarchy of wits – chimps and bonobos at the top with gorillas and orangs someway behind (and so in *Planet of the Apes* the gorillas served only as hired thugs). But those who work closely with gorillas and orangs say that they are just as thoughtful as chimps – although less obviously so because they are less extravert. Many monkeys are very bright, too. Even among humans, it can be very hard to judge who is really brighter than whom, or what this really means, and it's even harder with animals. There are so many confounding variables.

No-one doubts, though, that bonobos are second to none among all non-human animals; and no scientist has spent more time with them these past few decades than Sue Savage-Rumbaugh, now at the Great Ape Trust in Des Moines, Iowa. In her earlier job at Georgia State University Savage-Rumbaugh began by trying to teach a female called Matata to use a keyboard with geometric symbols. Matata didn't do too well but her son, Kanzi, apparently watched what was going on (although without seeming to) and he became the principal subject. He quickly learnt six symbols, then eighteen, and by 2003, when he was 23 years old, he had learnt 348. Some of the symbols are common nouns (yogurt, bowl, and so on) and some are verbs (chase, tickle, etcetera) while others are abstract (as in 'now', and 'bad'). By then, too, Kanzi also apparently knew the meaning of three thousand English spoken words.

Vocabulary, though, is only a part of verbal language. Equally vital (and in principle more so) is syntax, or grammar: the ability to re-arrange symbols to form new sentences. In humans this ability seems infinite – we can in principle express any idea that can be expressed in words. Bonobos seem more limited, but Kanzi nonetheless demonstrated some measure of 'proto-grammar'. Thus, out in the woods one day, Kanzi touched the symbols for 'marshmallow', and 'fire', and was duly given marshmallows and matches. He snapped some twigs and laid them out to make a fire, then lit them with the matches and toasted the marshmallows. (This is related by Paul Raffaele in the *Smithsonian* magazine of November 2006.)

In truth, this is even more impressive than it sounds. For in these experiments, Kanzi was very much playing away from home. Non-human animals do not habitually use verbal language. It's as if an English speaker was required to play a word-game in

Chinese without knowing the rules. Neither do bonobos or any other animal naturally live in laboratories. In the wild they live in forest, surrounded by their mostly familial social group. So when they take part in traditional scientific trials they are also on alien territory – as if we were being asked to perform while living among strangers and indeed semi-isolated, in a barracks. At the same time, in the confines of the laboratory, the subject animals have little or no opportunity to show their natural skills. Shakespeare would not have come out well in laboratory tests if no-one lent him a pen. All in all, as Dr Johnson said of dogs that attempt to walk on their hind legs, the wonder is not that bonobos and chimps generally fail to demonstrate advanced linguistic skills. The wonder is that they can do it at all.

Among the brightest non-primates are – wait for it! – spotted hyaenas. Over the years and centuries, hyaenas have had a most unfortunate press. They are ugly (or so many agree), they shamble, they feed on carrion, they crush the toughest bones in nature's most extraordinary jaws, and they snicker in a fashion that is deemed most sinister. Lions, in contrast, are billed as the kings of the jungle – brave and noble hunters and family men. But it is all most unjust. Hyaenas are formidable hunters too – and are no more drawn to carrion than lions are. Hyaenas too have a complex social life. Male lions leave most of the hunting to the females, and they kill, systematically, any cub that is not their own. They are, however, thought handsome, with a profile like that of a Victorian high-court judge, and of course they roar like any God.

In intelligence, though, hyaenas win hands down. Lions are rather dense – easily confused. If they weren't so enormously strong, they'd be a pushover. Hyaenas, on the other hand, are intellectuals. Second to none in his admiration for them is Simon Bearder, recently retired from Brookes University, Oxford, who has spent as much time as possible in Africa these past few decades, mainly studying bushbabies, but with an eye for other creatures too. He has tried, with fellow zoologists, to trap hyaenas. First he found that this is remarkably difficult. Secondly, he found it impossible to catch any individual hyaena in the same kind of trap more than once. They learn the mechanisms immediately. Furthermore, once a hyaena has been caught in any one kind of trap, it is impossible to catch any other individual in its same family group with the same kind of trap. Evidently, news of the perils of each design spreads within the group. When hyaenas are caught in traps, says Professor Bearder, they usually lie down in a resigned sort of way, and wait to be let out, as if to say in the manner of old-style villains, 'It's a fair cop, Guv!'. Only once, he says, has he seen a trapped hyaena get angry. Yet it seemed to be angry with itself: 'For God's sake! How could I have fallen for a trick like this?!' But then, this particular hyaena was the only one of all those caught by Simon and his colleagues that had had no chance to examine the mechanism – for the scientists held the door open on a long piece of string and let it go when he was inside. This, of course, is just an anecdote, but it makes the point.

Birds are excellent puzzle-solvers too – especially (or so it seems) parrots and crows. In the 1990s, carrion crows on a university campus in Japan, possibly stimulated by the general academic buzz, devised a variant, for cracking walnuts. For walnuts, at least when young, have a velvety skin around the shell and do not crack readily when dropped. So the crows, instead, put them in front of cars that had stopped for the lights. They placed them with the lights were red, retreated when the lights turned green, then returned when the lights turned red again to gather the spoils. If the nuts remained unbroken they moved them, to be more in line with the wheels. Later, crows in California learned the same trick. As always, it is possible to explain such behaviour away in behaviourist terms: by some lucky accident crows found that nuts dropped on the road got broken by cars, and so the behaviour was rewarded and 'reinforced', and so on and so on. But the crows certainly seem to have puzzled it out and there is no overwhelming reason, apart from prejudice, to suppose that they did not.

Star of crow puzzle-solvers, though, are the New Caledonian crows studied at Oxford University. Professor Alex Kacelnik and his colleagues gave them wire hooks with which to fish out little containers of food from long tubes. They did this readily enough, which is fairly impressive. Much more impressive, though, was a female called Betty. When a male flew off with the one remaining hook, she took a straight piece of wire and bent it, to make a hook of her own. Other birds have been shown to use tools, including some warblers which dig insects out of stems using thorns as probes. But Betty is the first known avian tool-maker. What do they do in the wild? Studies are now afoot to find out.

Some birds, too, are at least as good as non-human mammals are at dealing with words. Unlike non-human mammals, too, many are brilliant mimics and can imitate human speech – including parrots, crows, mynahs, starlings (which are related to mynahs), mocking birds, and lyre-birds. Ability to 'speak' is not directly related to overall brain power (some non-speakers are at least as adept in other fields) but some birds clearly use speech inventively and that surely means something. Best-known of all talking parrots was Alex the African Grey, who was studied for more than thirty years (1976 to 2007) by Irene Pepperberg, first at the University of Arizona, then at Harvard and Brandeis. By the time of his death Alex knew 150 words, including nouns, verbs, and adjectives. That is not a vast vocabulary – but, like the bonobo Kanzi, he could also to re-arrange words in ways that suggested he understood the underlying grammar. When he saw an apple for the first time he called it a 'banerry' – apparently combining 'banana' with 'cherry': two fruits he knew well.

Alex could also count up to a point and, said Dr Pepperberg, he had learnt the concept of 'zero' (which apparently the mathematicians of Ancient Rome never did). By parrot standards he was young when he died and we will never know what heights he might have reached. Already, though, there are potentially greater stars including the African Grey N'kisi, who by 2004 knew 950 words. N'kisi also had a sense of humour (and so do many other animals, according to those who know them well, including primates, dogs, and baby elephants). Of course, all these talking birds have their detractors. Do they really 'understand', or are they merely clever robots, learning (as a robot might easily do) to associate particular sounds with particular objects and actions? Such caveats should be taken seriously – but we should not just assume that the negative interpretation is the right one, just because it conforms more readily with existing theory. By the same token, as with all such experiments, we should remember that the animals are playing away from home. How much more might we see of their abilities if we spoke *their* language, and visited them in their own jungle homes?

Perhaps, though – or indeed probably – the ability of animals to solve problems is only an extension of their more general ability, and their ever-present need, to function as social beings.

Clever, social animals

Two biologists in the mid-twentieth century did more than anyone to break the stranglehold of Descartes and the behaviourists, and to restore the idea that animals are, indeed, thinking and sentient beings – truly, our fellow creatures. One was Jane Goodall, with her close-up, personalised studies of chimpanzees in Tanzania, beginning in the 1960s. The other was Konrad Lorenz who for some decades before and after World War II shared his family's Austrian *Schloss* with a wide variety of mammals and birds that he got to know as individuals, and he described his insights in the revelatory *King Solomon's Ring* in 1949 (published in English in 1952).

Jane Goodall stressed from the outset that chimps cannot be understood unless we empathise with them. They cannot adequately be explained as if they were robots – or at least they can (anything can) but the explanations rapidly become contrived to the point of nonsense. Lorenz obviously took the same view but clearly felt that he ought to conform to the standard idea of how scientists should operate and ended up in a kind of schizophrenia. So in *King Solomon's Ring* he tells us that Grey Lag Geese can be 'wonderfully affectionate' and 'long for human society' and that the ganders 'fall in love' – but he also warns us not get carried away, lest we fall into the trap of anthropomorphism. Most revealing, though, is his praise for the English author Rudyard Kipling and for the Swedish Selma Lagerlöf. Both wrote about animals in shamelessly anthropomorphic vein (and both won Nobel Prizes for literature) and in the introduction to *King Solomon's Ring* Lorenz writes:

> They may daringly let the animal speak like a human being, they may even ascribe human motives to its actions and yet ... they convey a true impression of what a wild animal is like.[19]

Indeed. But to be scientifically respectable in the mid-twentieth century the 'true impression' had to be couched in behaviourist-speak.

Of all Lorenz's stories, the one I like best is of the jackdaws that lived, semi-captive and semi-wild, on the roof. Jackdaws, like many animals that live in groups, have a strict sense of hierarchy. Unlike some other creatures, though, jackdaws like to establish the ranking peaceably:

> After some few disputes, which need not necessarily lead to blows, each bird knows which of the others she had to fear and which must show respect to her.[20]

So who decides who is boss? Myth, bolstered by a crude interpretation of Spencer's 'survival of the fittest', tells us that the biggest and strongest must rule. But animals in general are far more subtle than that. For, says Lorenz:

> Not only physical strength but also personal courage, energy, and even the self-assurance of every individual bird are decisive. *(Ibid.)*

The same has often been observed in primates: the alpha animals, male or female, are not necessarily the toughies, but are the ones that the other troop members approve of. In international human politics, might is generally deemed to be right, but our fellow creatures tend to be more discriminating. Just to divert: Rachel Hevesi describes a takeover of power among woolly monkeys at the Woolly Monkey Sanctuary in Poole, in the west of England, where she was curator. When the old boss monkey grew sick, his place was taken by his obvious successor – a big young extravert male. But the incomer was not up to the job. One day in a fit of pique he brushed an infant from his back – which, in woolly monkey society, is just not done. Woolly monkeys indulge their youngsters. Immediately the females shunned him – and as in many other primates societies it's the females' opinion that counts. His place was taken by another young male who was smaller but altogether kinder and more balanced. It took a long time for the bully to gain acceptance again (and of course, he never again became the number one).

In our own time, Dutch primatologist Frans de Waal has in many ways picked up the Lorenz mantle. The host of stories – some rigorous studies, and some more anecdotal that de Waal presents in *The Age of Empathy* – make four essential points. First, the behaviour of animals when they are given some freedom of action, and particularly their social behaviour, is far too complex to be sensibly explained away in the mechanistic terms of the behaviourists. Indeed, he says:

> ... a school of psychology that, in my opinion, has wreaked more havoc than any other: *behaviourism.*[21]

Secondly, individual animals in social groups are not constantly fighting for supremacy. Indeed, they spend a great deal of energy making sure that the group as a whole is peaceful and successful.

Thirdly, altruism (defined in non-moralistic terms – as behaviour that helps other individuals, even at cost to the helper) is common among animals of all kinds, and cannot always be explained away as kin selection. Non-relatives help each other and form alliances that can properly be called friendships.

Furthermore animals such as primates gain kudos within their groups by showing themselves to be kindly (as demonstrated by the Cornwall woolly monkeys.

Fourthly, empathy is *real.* It exists between human beings, of course; but also between humans and other animals, and between non-human animals – even between animals of different species. Empathy is not the same as sympathy, at least as the word 'sympathy' is commonly used. Sympathy can mean, in its crudest form, that you simply feel sorry for somebody else. Empathy means that you more or less literally feel their pain, or their joy; or at least it means that your own thoughts and emotions are in tune with theirs.

In short, we do not find that animals are robots blindly driven by hunger, fear, and lust. We find if we look sympathetically, and trust our own instincts, and shake off the science-rooted dogma which tells us that the universe and life and other species are just machines, we find that what Darwin said in *The Descent of Man* is true: that there is no human quality, including the most admirable of human qualities, that cannot be found in some form or other in other animals. There is no sharp cut-off between us and them.

Thus, as de Waal describes, plain, simple altruistic sociality is demonstrated by chimps studied in Tai National Park in Ivory Coast, which helped their companions to recover from leopard attack. They licked and cleaned their fellows' wounds, warded off the flies, and slowed their pace when they travelled so that the injured ones could keep up. Similarly, Jane Goodall tells the story of a chimp called Fifi and her two sons, Freud and Frodo. When Freud hurt his foot, it was Frodo who stopped and whimpered to tell Fifi not to race ahead; and when the somewhat forgetful Fifi did finally stop, Frodo would gaze at Freud's injured foot as if to communicate the problem. Frodo's behaviour isn't mere helpfulness. It implies true empathy – Frodo appreciated Freud's pain. It is also a fine example of communication in a wild animal – not merely of the individual's own state of mind, but of someone else's.

Some primates, says de Waal, 'invest in the community as a whole'. They don't just mill about in the group, interacting with the others right enough but basically looking after their own interests. They demonstrate true 'group oriented behaviour'.

Chinese golden monkeys are a case in point. They live in harems, with one big and colourful male, and several smaller and more modestly turned out females. A key role for the males is the keep the peace – for the females tend to be quarrelsome and when they fall out the male will stand between them, looking from one to the other with a friendly expression, often holding out their hands to each of them like a traffic cop ('Now take it easy there!'). With chimps, the sex roles are sometimes reversed. Males may threaten each other with weapons, whereupon the females disarm them – which the males allow them to do, even though the males are much bigger and stronger. They are like drunks in the pub car-park, half Dutch courage and half funk – 'Hold me back, somebody!'

Modern biologists, observing animals up close in the wild and under various degrees of constraint have now provided hundreds of such examples. The point is already established well beyond all reasonable doubt that Descartes got it wrong – animals emphatically are not automata – and that Darwin got it right: it is very hard indeed to identify human mental abilities that do not have their parallel in other animals. Indeed there are very good reasons to think that other animals have abilities that may be far beyond our own (or at least are far beyond what is normally apparent) as outlined in particular by Rupert Sheldrake in books such as *Dogs that know when their owners are coming home* and, most recently, *The Science Delusion.* The past few decades have seen a radical transformation in general understanding, from strictly mechanistic behaviourism and absolute condemnation of anthropomorphism to the notion summarised by Pat Bateson – that we cannot hope to understand animals *without* comparing them to ourselves. In the years to come we may yet see a further transformation: perhaps the widespread acceptance that mental capacities that are now regarded with amused contempt in most scientific circles, including what is commonly known as telepathy, should be taken very seriously indeed. Indeed, as we will see later, in the ancient idea that consciousness is not simply a property of brains, but is a property of the universe, is also rapidly gaining ground. It may well be, indeed, that by the middle of this century, the mechanistic view of the 'higher' mental faculties that is still the orthodoxy, will seem very old hat indeed; at least as primitive as Descartes' speculations seem now.

On a different note, there is clearly a relationship between the capacity of animals, including human animals, to be social and unselfish, and the concept of morality. Sociality and cooperativeness cannot by themselves be called moral, because the concept of morality implies a conscious decision to behave morally. Honey-bees, therefore, cannot properly be considered moral (despite the comments of T.H. Huxley) because they don't, we might reasonably assume, *choose* to behave morally. They really do (we might reasonably assume) go where their genes lead them. But sociality nonetheless *underpins* morality. It is the ground condition. Thus our innate, gene-directed biology inclines us towards unselfishness and hence to morality – much more than it inclines us to be selfish and solipsistic. Basically, in short, we and other animals are *nice*. In our own interests, and in the interests of truth, we should not allow ourselves to be persuaded otherwise, as Enlightenment thinkers and some politicians choose to suggest.

But we will come to this. In the next chapter I want to show that the cooperativeness that is seen within species, and sometimes between intelligent animals of different species, extends through all of nature. Competition is a fact. But cooperativeness is the essence.

5. Gaia

It's not too surprising to find that creatures of the same kind often cooperate. Clearly they or their genes may benefit in all kinds of ways if they help each other. It's more surprising to find cooperation on the grand scale: that all of nature, allegedly so 'red in tooth and claw', operates overall as one giant cooperative. Yet if it did not, it could not function, and all would die. Indeed, one of the great insights of the twentieth century is that of the English scientist James Lovelock, who has suggested that the whole Earth and all its creatures together have all the qualities not simply of an alliance but of a living organism. Lovelock's neighbour and friend, the novelist William Golding, suggested that this hypothetical whole-Earth organism should be called 'Gaia', after the Greek goddess of the Earth. The 'Gaia hypothesis' is surely one of the great insights of the twentieth century, to rank with those of Jane Goodall on chimps, Lynn Margulis on symbiosis and the nature of the eukaryotic cell, and Karl Woese on the domains of living creatures.

To be sure, it isn't quite accurate to say that Gaia *is* an organism. But it is analogous to an organism: the living world and the non-living world of rock and air and water together display the principal characteristics of an organism. Most obviously, like an organism, Gaia is infinitely complex – and all its different components interrelate, and all are ultimately dependent on the others. But the relationships between the many players are typically 'non-linear' – far from obvious or straightforward; and hence, in detail, they are innately unpredictable. Most extraordinarily – the fact has become clear only over the past few decades – this all-embracing Gaia, like any living creature, has the quality of homeostasis: it maintains its own internal environment (the '*milieu intérieur*') in a state that fosters its own survival.

We should look at the qualities of nature – of Gaia – one by one: its complexity; its interrelatedness; its non-linearity; its capacity for homeostasis; and its unpredictability and ultimate unknowability.

Numbers

We can express Gaia's complexity in all kinds of ways – not least by the number of species. But we see very soon how ultimately forlorn such endeavour really is. 'How many species are out there?' seems the most elementary of questions – the kind a child might ask. But although scientists have been making formal lists for several hundred years, and keen and competent naturalists have been collecting all that breathes at least since Aristotle, and hunters and gatherers and farmers well before the Greeks had extraordinary knowledge of their non-human neighbours, we still don't know to within an order of magnitude what's out there. Still less do we know what lived in the past, and since all creatures are evolved, what lived in the past is pertinent.

It's hard, for a start, even to do what seems like the simplest thing: just to identify and tot up whatever is out there. Nothing, for example, is more conspicuous than a tree. It wouldn't take a botanist long to count the number of different species in the average piece of pristine British woodland (not that there is much pristine woodland left). Britain as a whole has only thirty-nine native species of tree – after years of intensive exploration we can be pretty sure of that – and no one wood, if it's not full of exotics, is likely to contain more than about half a dozen. Tropical rainforest, you might suppose, is just the same problem scaled up.

Well – up to a point. But scaling up makes all the difference. For while temperate forests like Britain's contain relatively few species, tropical forests typically have thousands: there may be three hundred different kinds in a single hectare. It's easy to get to grips with our natives (and with the commoner imported exotics) and to recognise them all at glance, winter or summer, but when there are hundreds within eyeshot and most are unfamiliar it's far from easy.

In fact, although quite a lot of people have spent quite a lot of time on it (not enough, in either case, for all kinds of reasons) we really have only a hazy idea of how many trees live in the tropics; and since we know that *most* of the world's species, by far, do live in the tropics, we have little idea of how many there are in all. The number of different species of trees in the Neotropics (the American tropics, from northern Mexico down to northern Chile and Argentina) has been estimated at thirty thousand; and there could be sixty thousand or so different tree species in the world as a whole. But although these are informed guesses, they are guesses nonetheless.

Just how shockingly little we know of what's out there and who's who was illustrated in the 1970s by another Smithsonian biologist, Terry Erwin. He 'knocked down' (meaning anaesthetised, which in effect means killed) all the small creatures living in a few specimens in the crowns of just one kind of tree in Panama, and gathered them all up. There were too many different types to count exhaustively so he focused on the beetles (though he left out the weevils, though they are beetles too). This was reasonable. Beetles are easy to identify and they account for about half of all known species of animals so if you know the beetles you can easily estimate the rest.

In the crowns alone (he had to make a separate estimate of the number in the roots) Erwin found 1100 different species of beetle, some of them unknown to science. So how many in the forest as a whole? If the same beetles lived on every kind of tree, then there would only be 1100 overall (which of course is known to be untrue because many thousands have been identified). But if each tree had its own exclusive suite of beetle, and all of them had at least 1000 different types, then the total of all beetle species in the neotropical forest would be 1100 x 30,000 (the number of trees) which is 33 million. Both extremes seem highly unlikely –

obviously some beetles are exclusive to particular trees but others live on several or many different types – so the true number presumably is somewhere between 1100 and 30 million.

In the end, after making reasonable assumptions about the total number of species of all kinds of creature relative to beetles, and about the proportion of all world species that live in the neotropics, and so on and so on, Erwin suggested that the total number of species of all creatures in the world as a whole was possibly around thirty million, but could be as high as one hundred million. Since fewer than two million species of all types had and have been described, biologists in general were shocked by Erwin's conclusions, not to say stunned. But although his argument clearly depended on a great many suppositions, none could find serious fault with his reasoning. Since then, for a great many reasons, including habitual cautiousness, Erwin's guess has been modified, and many now feel that the true number of species on Earth right now is probably about five to eight million. But that too is a guess; and even if we bend over backwards to be nice to ourselves, it's clear that we don't yet know the half.

But that's not the end of it. Most biologists, when they think of wild ecosystems, tend to focus on the visible creatures: animals, plants, and fungi. But most of the world's biomass is microbial: bacteria and the archaeans (as outlined in Chapter 3). About fifty thousand species of microbe have been described, although it is hard to say exactly what a bacterial or archaean species really is, since they are so good at swapping DNA wholesale, which significantly blurs the boundaries between them.

However, as Norm Pace, now at Colorado university, has pointed out, bacteria traditionally were identified by putting bits of substrate – soil or cheese or whatever – in culture, and seeing what grew. This means, however, that the final catalogue depends not so much on how many there really are in the substrate, but on what proportion of them are content to grow in the medium provided. So Professor Pace decided to adopt a more direct method. He used DNA probes to reveal, not the bacterial organisms themselves, but their DNA. Even more shockingly than Terry Erwin, he showed that of all the bacterial DNA present in soil, only one ten thousandth of it belongs to bacterial species that have been previously identified. Thus the true number of bacterial species could be 50,000 times 10,000 = 500 million. This may seem like a ridiculously huge number but still we have to say – who knows? Whatever the number may be, it's clear that our present knowledge falls far short.

Can we fill the knowledge gap with more research? Up to a point, perhaps – but not a very impressive point. To describe adequately all the creatures that we might find in nature would surely occupy the entire human population for hundreds of years to come – assuming that we were all expert taxonomists. Since the number of professional taxonomists is diminishing (nobody pays for science these days unless there's immediate profit in it) and there were never very many to begin with, we can safely say that for practical reasons alone the task will never be completed.

But there are theoretical reasons too why the task cannot be completed. As the English philosopher J.S. Mill pointed out in a general context in the mid-nineteenth century, however much we know or think we know about anything, we can never be sure that we haven't missed something. Furthermore, it is logically impossible to know how much we don't know unless we are already omniscient. So however vast the world's inventory of species may become, we can be sure only that it is incomplete.

Right now, though, for all the heroic efforts of the past few thousand years we can be pretty sure too that whatever we know or think we know only a fraction and certainly far less than half of what's out there.

So much for the ludicrous boast we have so often heard from on high that scientists now 'understand' nature, and can control or even 'conquer' it. People who believe such things, or act as if they do, are seriously deluded and must be treated with extreme caution.

But for the science of ecology, the inventory of species is merely the cast-list. What counts is the play itself – the way the players interact; and the number of possible interactions, even in an apparently simple ecosystem, is very large indeed, and in principle infinite.

Interactions

No man is an island, said John Donne; and what is true of each and every human being is true of all creatures. All exist in relationships, and indeed they cannot live outside relationships. Relationships by definition always involve more than one individual and in practice the more we look the more we see that all are linked to all.

Ecology is the study of nature's relationships and its scope is infinite, as can be seen even from the crudest arithmetic. The number of possible relationships depends to some extent on the number of players – both on the number of species and on the range of variation within those species, for the youngsters and oldsters within any one species may form quite different relationships and have a quite different ecological significance. Each individual creature throughout its life may interact directly with many different creatures of different kinds.

Overall, the number of possible relationships between all the different creatures in any one ecosystem is the number of *permutations* possible, between all the different creatures present. But the impact of any one species on any one ecosystem depends not only on what kind of creature it is, but also on how many there are – and the population of any one species can range from almost zero to the vast majority. Given that there is such variation, the number of possible permutations within any one community of creatures in any one ecosystem is not simply very large, but is infinite. Ecology has become a fabulous and in some ways precise science but in the end its task is impossible. In practice, ecology is mostly a combination of natural history – just trying to see as best we can what actually happens in nature – and rarefied maths: trying to reduce very large numbers to manageable formulae that don't provide a complete insight or anything like, but do at least give a handle on what's going on (or that at least is the hope).

A few anecdotes – they can be no more than that – will make the point.

The absolute interdependence of plants, fungi, and bacteria

Take, for example, the many relationships between plants and fungi. Both groups evidently originated in water, and found their way on to land some hundreds of millions of years ago. They are very different kinds of organism – surprisingly, fungi are far more closely related to animals than to plants. But to judge from the range and intimacy of their present relationships it is possible that neither kingdom would have succeeded on land without the other. Wherever we look we find plants and fungi locked in symbiosis, and although some plants can sometimes grow in a fungus-free environment (and if they could not, the horticultural technique of hydroponics would be impossible) most wild plants rely on their relationships with fungi and many, including most trees, rely on fungi absolutely.

The most visible plant-fungus symbiosis – you see examples on every tree trunk and rock all over the world when the air is not too polluted – is that of lichens; and although no-one knows when the very first ones appeared it's fair to guess that they are ancient, given that they involve algae, which are among the most primitive *bona fide* plants of all. (I say *'bona fide'* plants because there are many small green organisms that are not, strictly speaking, plants.) The lichens demonstrate true 'mutualism'. The fungi provide anchorage while they forage for nutrients in unpromising films of moisture and in the fabric of the substrate while the algae, safely lodged on the fungal base, provide sugars by photosynthesis. About thirty thousand kinds of lichen are known worldwide. Locked together they thrive in the most meagre conditions, down to and including the tombstones in windswept

Yorkshire churchyards, where no other kind of multi-cellular organism – even mosses – could find a niche. Lichens are the great pioneers. They gather debris as they grow and thus provide the rudiments of soil – and then the mosses can take hold and after that the floodgates are open to whatever is green.

Less obvious to the eye but of incomparable importance to the whole world are mycorrhizae. Here the fungi grow in intimate relationship with the roots of plants. Their hyphae enwrap and in many cases penetrate the cell roots – and then spread far beyond the range of the roots themselves. The physiological arrangement is essentially the same as with lichens: the fungal hyphae penetrate spaces where the roots cannot, which is a physical extension; and they secrete enzymes to break down organic matter in the soil, and to prise phosphate from rock, which plant roots cannot do. As in lichens, the plants in exchange provide sugars by photosynthesis. Green plants derive their energy from the sun and take carbon from the air and water and minerals from the ground and so they are 'self-feeding' – and are known technically as 'autotrophs'. Fungi and animals are 'heterotrophs' – feeding on organic matter that has already been made by some other creature (and ultimately by some autotroph). The plant and mycorrhiza together are both autotrophic and heterotrophic and so the plant-mycorrhiza complex gets the best of all worlds.

The hyphae from the mycorrhizae of neighbouring trees join up, and so the whole forest may form one giant consortium, with different regions of the grand alliance making a special contribution of whatever nutrients are abundant in their own particular spot. The complexity can be enormous. Any one tree may have up to three hundred different mycorrhizal species in its roots; and the total number of mycorrhizal species found in any one species – as shown not least by studies of Douglas firs – may be around two thousand. Douglas firs are in the pine family, Pinaceae; and the Pinaceae as a whole are absolutely reliant on mycorrhizae, and make such good use of them that between them they grow everywhere from the most extreme north to what often look like sand-dunes in southern Spain and North Africa.

Without mycorrhizae they would have no chance. *Most* wild plants rely heavily on mycorrhizae. Just a few can do without them, and they can be seen as specialists which grow very well on freshly-broken ground where the mycorrhizae have been broken up and largely destroyed. Such plants are known as 'pioneers' and are not too common in nature, because freshly-broken ground in nature is not too common. The soil quickly settles and the mycorrhizae return. But farmers commonly see it as their job to break soil. The pioneers which then invade are known as weeds – and in agriculture, of course, are extremely common. There is much interest now in perennial crops that grow on unbroken soil in which mycorrhizae flourish and the weedy pioneers are kept in their place.

All plants too in a state of nature rely absolutely on bacteria. Many simply live in the soil and, like fungi, break down organic material into its component ions which the plant can absorb as nutrients. But some do more than this. They 'fix' nitrogen: turning inert, atmospheric, molecular nitrogen gas into ammonium ions which plants can absorb, while other bacteria help to turn the ammonium into nitrate which is also absorbed. 'Modern' industrial farmers feed their crops with fertilisers commonly marked 'NPK' – supplying the three essential macro-nutrients of nitrogen, phosphorus, and potassium. Nitrogen-fixing bacteria provide N fertiliser for free, mega-tons of it every day. Wild nature and organic farming rely absolutely on nitrogen-fixing bacteria. Wild nature is super-abundant and organic farming could feed the world (even though agroindustrialists and the governments that serve them deny this).

Plants of many different families form special relationships with various nitrogen-fixing bacteria. Many grasses, for example, including maize, are known to secrete organic materials which favour free-living nitrogen-fixing bacteria which thus flourish around their roots. Some other families have nodules in their roots which provide special board and lodging for nitrogen-fixers. Best known and most significant of this are the many different members of the family Fabaceae, formerly known as Leguminosae, and hence still known colloquially as 'legumes'. They house N-fixing bacteria of the genus *Rhizobium* within their root nodules. Legumes include peas and beans among the crop plants, including the hugely significant cow-peas, pigeon-peas, and chick-peas of the tropics; clovers and vetches – essential sources of fertility in traditional grazing systems; a range of shrubs including gorse which thus can grow on seriously impoverished soil; and, perhaps most importantly of all, a huge variety of trees. Most significant among the trees are the many acacias of the tropics and subtropics. But a high proportion of the trees of tropical forests are legumes too. They are essential to the wellbeing of the whole forest. Tropical rainforests are the most diverse environments on Earth and prime regulators of global climate – removing carbon dioxide and circulating water. But it all depends in the end on the symbiotic relationship between trees, often giant trees, and their tiny bacterial lodgers.

Yet the legumes are not the only flowering plants to form such relationships. The alders enjoy a similar relationship with filamentous bacteria of the genus *Frankia*; and the various species of alder are extremely significant, mainly in temperate places where for whatever reason the soil is low in nitrogen, not least on dank muddy river-banks. In some areas, as in some forests in Latvia, alders dominate and are the principal timber – everything from barn-doors to pianos.

The nitrogen fixing trees and other plants also maintain their relationships with mycorrhizal fungus. Each forest tree, in short, is an ecosystem in its own right. Or, we might say, each is a mini-Gaia. Ecologists are sometimes wont to say that the boreal (extreme northern) forests of Canada and Siberia are 'simple', because they are dominated by a mere nine species of tree, including eight conifers and the deciduous quaking aspen. But beneath the ground, where innumerable bacteria and fungi work their magic, the boreal ecosystem may well be as complex as a whole tropical forest.

The same autotroph-heterotroph best-of-both-worlds relationship is repeated many times, with many different combinations of species, in the sea and in freshwater. Giant clams with the green protists known as dinoflagellates; hydras with the single-celled algae, *Chlorella.* Corals die without their cargos of zooxanthellae – more single-celled protists. Coral reefs, like forests, are monumental; and, like forest trees, they contrive to be heterotrophic and autotrophic at the same time. All such symbioses reflect the kind of arrangements that first gave rise to eukaryotic cells – when bacterial lodgers became so closely integrated with their hosts that they exchanged genes and became one organism. Of course, most organelles arose as bacteria – but some, it seems, arose as full fledged eukaryotes in their own right. Again the message comes loud and clear: cooperation is what really counts in nature.

But let's return to the forest.

Plants and their animal helpers

When trees have grown, with the help of their fungal and microbial chums, they seek to reproduce. Their flowers (or male catkins) produce pollen, from the anthers, which contain the male gamete: and the pollen then finds its way by various means to the receptive female stigmas, which lead to the ovary, which contains the female gametes. But trees, most emphatically, are rooted to the spot – so what provides the transport? Many temperate trees including most conifers rely on wind; there are plenty of gentle breezes in high latitudes and open woods, and since the total diversity of trees in temperate zones is low, there's a fair chance that pollen from any one tree will find its way to another tree of the same species. But breezes don't function so well in equatorial forest so most tropical trees rely on animals to carry their pollen. Pollinators are of many kinds, including lemurs and arboreal marsupials and bats and humming birds and many more. But insects are the chief pollinators: many beetles (which were probably the first ever insect pollinators) but mainly Hymenoptera, including bees of many kinds, and Lepidoptera, butterflies (by day) and moths (by night). Then when the ovules are fertilised and grown into seeds the seeds, or the fruits that contain them, need to be dispersed. Many temperate trees and most conifers again rely on the wind but again, the principal seed-dispersers of the tropics are various animals.

The relationship between flowering plants and their various pollinators and seed dispersers is so intimate and so vital to both that all the creatures involved have shaped each others' evolution. Indeed, plant-animal relationships are prime examples of what Darwin called 'co-evolution'. Since most trees, and most plants of all kinds are tropical, it's clear that *most* plants rely on animals for their own reproduction; and if organisms fail to reproduce, their lineages come to a halt. By the same token, the extensive orders of the Hymenoptera and Lepidoptera would hardly have a presence on this Earth at all were it not for the plants they have evolved alongside. Flowering plants are their *raison d'être.* Overall, then, the stunning diversity of nature is driven in very large part by the central need for mutualistic (mutually beneficial) co-evolution.

For once a mutualistic relationship has arisen, further complexity often follows. My favourite example is of a tree that I was introduced to about a decade ago on Barro Colorado in Panama, by Egbert Leigh of the Smithsonian Tropical Research Institute (STRI). This is *Dipterix panamensis,* the almendro – a stunning legume, with a hemispherical crown and salmon-pink bark, and compound laburnum-like leaves spiralled around the twigs. As a legume the almendro of course relies on its resident N-fixing rhizobia, and like all trees it has its own cortège of mycorrhizae.

In April and early May in Panama, the rains come, and in June to July, evidently triggered by the rains, the almendro sends out bunches of pink flowers. Like most tropical trees, individual almendros are widely spaced – there is only about one individual per hectare in Barro Colorado – but bees provide the necessary pollination. A few weeks later the fruits appear – hard wooden pods coated with a thin layer of sweet pulp, each with a single seed inside: about twenty per square metre in a good year.

When the fruits emerge, says Dr Leigh, 'swarms of animals come to the feast'. Some of the visitors take the fruit directly from the tree, including various monkeys and members of the mammalian order of the Carnivora including the raccoon-like (but unrelated) coatis and the kinkajou – a strange member of the civet family with a prehensile tail like a monkey. (Carnivora of course are supposed to eat meat, but many in practice are omnivores including foxes, raccoons, and bears.) Squirrels abound and so do fruit bats. Some of the beneficiaries gather the fallen pods from the ground, including rodents such as the spiny rats, and agoutis and pacas, big relatives of the guinea-pig; and hoofed animals such as peccaries (New World pigs) and the occasional tapir. Of all the feasters, some just eat the sweet pulp around the pod while others gnaw through to the bean, including peccaries, squirrels, spiny rats, and agoutis.

Most of the feasters do the almendro no good at all – notably the squirrels, which simply eat the fruits off the tree and drop the debris straight to the ground. Monkeys do slightly better because although they eat a great deal they also carry some away from the tree and so do some dispersing – which is vital because young almendros, as is commonly the way with tropical trees, will not grow close to their parent.

Most important to the almendro, however, are the fruit bats. They do not eat the fruits *in situ*. They carry them away from the tree to eat in peace some distance away. This seems to be a survival tactic. They forage at night, and if they did hang around in the tree they would be ambushed by owls. Birds of prey in general are bad news for bats. Furthermore, the bats do not chew through the hard pods. They merely nibble away at the pulp, then drop the intact pod to the ground.

Whereupon the second true friends of the almendro take a hand – the agoutis. To be sure, they eat most of the pods with the seeds inside but they also bury some for later, just as squirrels bury nuts. In a bad fruiting year they eat all the pods and seeds. But in a good fruiting year there are too many to eat all at once and many are buried.

Still, though, the almendro is not out of the woods, metaphorically speaking, for the agoutis come back later to raid their stores. Even if they do forget some, and the seeds grow, they are likely to return to eat the seedlings.

Enter now the third essential player: the ocelot. Ocelots are small and beautiful spotted cats – not the kinds of creatures you would think of as symbionts of forest trees. Their role, though, in this elaborate symbiosis is to eat the agoutis. If they eat an appropriate number of agoutis after they have buried the almendro pods, and before they dig them up again, then some of the seeds will make it.

Clearly, the whole process is touch-and-go – but for many thousands of years, and possibly for millions, it has worked. Now, though, says Egbert Leigh, it doesn't seem to be working. Now there are not enough young trees to replace their ageing parents. One reason, it seems, is the loss of the ocelot. Carnivores in general are having a bad time in the modern world and the ones with the most attractive coats fare worst of all. There are laws to protect wild animals these days but wild animal trading nonetheless is one of the world's biggest export markets, second only to drugs and arms. Global warming is taking its toll, too. If almendros produce only a small number of fruits at a time then they will all get eaten. But they produce big crops only if the April rains fall sharp and hard. A half-hearted rainy season – stop-and-go, and not much – fails to stir the almendros into full reproductive mode. Global warming is blurring the seasons. Bumper crops are becoming rarer.

It took several decades of at times obsessive observation to uncover the almendro story. Yet the almendro is only one of an estimated thirty thousand trees in the American tropics. They all have stories to tell – and some will be even more complex. Some, when you trace them through, are as intricate and with as many sub-plots as *Middlemarch*. Yet as we have remarked the number of possible stories is not dependent only on the number of species but on the total number of possible relationships between them – which is a matter of permutation and is infinite. To study everything that could be found out as thoroughly as Egbert Leigh has studied the almendro would take more time than the human species has yet existed but still there would be more to do.

The almendro story tells us too – and so do a million other stories that might be told – that relationships in nature are, as the physicists say, *non-linear.* We don't see simple cause and effect, as we do in Isaac Newton's laws of mechanics. Any one event in nature may trigger many more, which trigger many more, and everything that happens interacts with everything else, and for a variety of reasons the outcome of each individual interaction cannot be precisely predicted. Non-linearity is a key concept, and so it becomes impossible, both in practice and in theory, to predict in any degree of detail how any intervention in a natural system will turn out. In the end, nature is beyond our ken. The idea that we can control or 'conquer' nature is not only supremely arrogant. It is supremely stupid. No-one remotely versed in the philosophy of science could employ such rhetoric, except in a spirit of extreme cynicism. The fact that people in high places do make such claims and act on them, and that their endeavours are supported by compliant intellectuals, serves only to show that we have allowed ourselves to be led and advised by supremely stupid and cynical people. More later.

Yet the world is even more strange and wonderful than traditional natural history and ecology can reveal. For when we look closely we see that the Earth as a whole is homeostatic; not a mere habitat for living creatures, but an organism in its own right; and one of which all living creatures, including ourselves, are a part. This is the true key to the Gaia hypothesis; what moves it beyond conventional ecology.

Homeostatic Earth: the Gaia Hypothesis

James Lovelock came into science by an unorthodox route and has never truly been a part of the standard academic establishment. He began by reading chemistry at Manchester University before

World War II but could afford only two years of the three-year course. He clearly made an impression however – for then he was recommended for a war-time post with Britain's Medical Research Council to work on burns. So in 1948 he was awarded a PhD by the London School of Hygiene and Tropical Medicine and went on to do research at Yale, the Baylor College of Medicine, and then Harvard.

The path that led Lovelock to the Gaia hypothesis properly began when he went to work for NASA in 1961 – not to study the Earth, but to devise instruments to look at the atmosphere and the surface of Mars. NASA sought in a general way to find out what Mars is really like. But also, more specifically, they wanted to address the most intriguing question of all: is there, or has there ever been, life on Mars? Research in the sixties was carried out from a distance, although the Mariner probes went close: studying for example the light coming from the Martian atmosphere, which can tell us a great deal about its composition. Later, in the 1970s, NASA launched the Viking probes which actually landed on Mars. Data from all those explorations showed beyond reasonable doubt that there isn't life on Mars but on the very day that I am writing this (August 5, 2012) NASA is about to land a substantial roving laboratory on Mars to explore its ancient rocks to see if there are any signs of life in times long past. The new Mars probe has the pleasant soubriquet of *Curiosity*.

But NASA's Martian studies are significant mainly for what they tell us about the Earth, and especially for what they have already told us about our own atmosphere. The Earth sciences were transformed in the 1960s and '70s, with Lovelock very much part of the transformation, and Mars provided the comparative model, the perfect complement to the Earthbound studies. This is very necessary, for as Rudyard Kipling observed in a slightly more parochial context, 'And what should they know of England who only England know?'

So what was the nature of this transformation, and how did the Martian studies contribute to it?

A conceptual shift

When I first studied chemistry at school in the mid 1950s we were told, and solemnly recorded in our shiny new notebooks, that the atmosphere contains 79% nitrogen and 21% oxygen with very small but important quantities of carbon dioxide (about 300 parts per million in those days), plus a short catalogue of 'rare' or 'noble' gases such as neon and argon, which just hung about the place, keeping themselves to themselves, and a variable quantity of water vapour. Nothing else was mentioned, as I remember. Ammonia, methane, various oxides of nitrogen and sulphur and so on, did not get a look in. I remember our first chemistry teacher, a former WWII munitions worker, treating some hapless child's suggestion that the air contained hydrogen with a literal snort of contempt, in the manner of an elephant seal. If the atmosphere did contain hydrogen, that ancient sage declared, it would explode. Hmm.

He told us though that oxygen is lively stuff, and carbon dioxide matters because we breathe it out and plants take it up and combine it with hydrogen which they prise from water, to make sugars, and thence to make proteins and fats and the other stuff of flesh, which we and other animals then eat. But nitrogen, the bulk of the air, was for all intents and purposes 'inert' – as, of course, were the self-satisfied noble gases. Overall, it seemed, apart from the oxygen-carbon dioxide cycle, the air was really just a 'medium'; a matrix, filling in the space between the ground and the sky. We were taught such stuff partly because we were only eleven at the time, and words like 'nitrogen' and 'oxygen' were fairly new to us, without burdening our brains any further. But also, at least in caricature, that was roughly how the atmosphere was widely understood in those days; essentially passive, like bathwater.

The same kind of principle applied to the sea and the world's fresh waters. They are media for fish to swim in and although they have interesting and complex chemistry, essentially they just *are*. So too with land; just rock, crumbled at the top and bound by organic débris to form soil. The three great media – land, sea, air – were seen to be separate, with no significant interchange between them. We were, though, given to understand that rivers make the sea salty by loading it with salts dissolved from the land (a common sense concept that has now been greatly modified, as we will see). Crucially, apart from the oxygen-carbon dioxide story (oxygen breathed in, carbon dioxide breathed out and taken up by plants, diatoms, and cyanobacteria) there was no significant traffic between living systems and the fabric of the Earth itself – rock, water, air. We creatures, it was assumed, just live here; taking what we need; discarding what we don't need; building some impressive edifices, (coral reefs; the dams of beavers; Paris) and making various kinds of mess; but otherwise having no great impact.

Even without any reference to Gaia, the general perception of mother Earth has changed radically over the past half century or so. The key word now is 'dynamic'. The various components of land, sea, and – on a longer time-scale but no less significantly – of rock, are in constant flux. The air is a mixture of gases of which some including nitrogen are relatively inert and so just sit there for long periods apparently doing very little. But the atmosphere also contains a host of 'trace' gases that are present in low quantities only because they are very reactive indeed so that as soon as they are released they react with something else. These traces include the kinds we didn't mention as junior chemists: methane (100 ppm, or parts per million), ammonia (1 ppm), nitrous oxide (10 ppm), various sulphurous gases (0.01 ppm) and a range of organic compounds that include methyl chloride (0.1 ppm) and methyl iodide (0.0001 ppm).

Because these trace compounds are so reactive they are very important indeed, and in many different ways. So the *amount* of a given gas in the atmosphere (or of a particular salt in the sea) bears very little relationship to its overall significance. It's also clear now that all the chemical entities interact with all the others – so the atmosphere and the sea, together with the rocks of the land and the stirring mantle beneath, form one great chemical ecosystem, as intricate on the atomic level as the living ecosystem of a tropical forest. The whole potent mix is affected in various ways by solar radiation of various wavelengths – and it's a two way interaction: the chemistry of the atmosphere may greatly reduce or sometimes enhance the penetrating power of the radiation and so affect all the rest. In general, too, real chemists do not see chemical reactions in the way that we were taught as schoolboys – as cumbersome collisions of atoms and molecules, conforming to the simplified equations of textbooks. Instead they envisage the world as clouds of electrons and ions (atoms with outer electrons added to or taken away) and between one end of a textbook equation and the other there may be a whole chain of very short-lived intermediaries, many of which may catalyse other reactions elsewhere. Earth, air, and sea are not separate. There is enormous and rapid exchange between the three, with everything affecting everything else.

James Lovelock helped in various ways to paint this modern picture, to show how intricate it is. But he added one extra, crucial insight. For he saw, as others had not, that living creatures aren't just passengers on this Earth. They – we – are key players. The chemistry of the atmosphere, the seas, and the underlying rock is as it is largely because we have helped to make it so. Living things are very much a part of the great chemical exchange. At first many people, including many scientists, found this incredible, and some still do. Many American politicians and industrialists continue to question whether the activities of human beings, just one species among many, can *really* be having such a huge effect on carbon dioxide and methane levels now generally suggested, and so upsetting the global climate. Some of them are questioning this because they have a vested interest in the status quo but others are genuinely incredulous. Yet their incredulity is misplaced. The overall evidence now seems incontrovertible: the chemistry, the climate, and the physical structure of the whole Earth are profoundly influenced by the presence of life. Without life, our planet would be a hideous, roasting rock with an atmosphere that would be lethal in seconds to almost any present-day creature; and its clear that living creatures, now including ourselves, have been exerting our tremendous influence since life began, at least 3.5 billion years ago – through most of the history of the Earth itself, in fact.

Yet there is one more twist, with huge implications both scientific and metaphysical. For, as Lovelock saw early on, the changes wrought by living creatures to the make-up of the Earth are of a kind that make it easier – or indeed make it possible – to live. Life makes the changes that are good for life. That is the essence of homeostasis. The world is indeed an organism, and Gaia is indeed an appropriate name for it.

All this is an absolute conceptual transformation from the innocent days even of the mid twentieth century, when the Earth at least of elementary textbooks was just rock, and air, and water and we and our fellow creatures were mere campers. Now the world is rightly seen as one vast integrated system. Yet John Donne's adage – 'No man is an island' – applies just as well to Gaia as to ourselves or to any of Gaia's creatures. For all earthly life depends on solar radiation – the right intensity, the right wavelengths; and the tides have a huge impact on the ecology of the world, and they of course are driven by the moon. The whole solar system is our habitat.

How do we know whether life is present or not?

NASA's studies of Mars in the 1960s and 1970s encouraged thoughts of Gaia in various ways – not least by raising the fundamental question, how would we know life if we saw it? And if we didn't find it, how could we tell if there had been life in the past, yet now extinct? How, if no little green men appeared, with TV aerials in their heads and the customary, 'Greetings, Earthling!'?

No-one at NASA seriously expected little green men but they did think it reasonable to look for simple organisms and, even more realistically, to seek out the kind of molecules that we associate with life, such as amino acids (the stuff of proteins) and nucleotides (the stuff of nucleic acids). It also seemed reasonable to look for the kind of conditions that could support life, such as water; or for signs that such conditions had existed in the past (for example for the patterns of erosion that water causes). Of course, all such investigation was based on a preconception of what life in general is really like and what it needs – a preconception based on our knowledge of life on Earth.

Lovelock, though, conceived a broader approach – based on the physicist's grand notion of entropy, as well as our knowledge of life on Earth. The principle of entropy says that everything that exists in the universe tends to find its own state of least energy; or, to put it another way, that systems that are orderly (which in general require energy to keep them together) tend to become disorderly. So it is that houses fall down unless they are actively maintained, meaning that energy is put into them to keep them intact. Explosive chemicals explode: when suitably triggered their high-energy molecules break down into relatively stable compounds such as carbon dioxide and oxides of nitrogen, and the energy thus released makes mayhem elsewhere. Thus bombs reduce houses, which need inputs of energy to keep them up, to rubble, which doesn't. On the grander scale, stars run their course and die. In general, then, entropy is a measure of disorderliness, or of lack of energy. Over time, unless actively inhibited, all the components of the universe and ultimately the universe itself must slide towards a state of maximum entropy. This at least is the standard account of things.

But life seems to defy entropy. Living systems are wonderfully complex and wonderfully orderly: and as they grow from egg to adult their entropy if anything diminishes – they impress more disorderly stuff into their orderly selves. Some scientists and philosophers have seen this as a profound puzzle for life at first sight seems to defy one of the most fundamental principles of physics, and that seems to make it magical. But nothing breaks the fundamental principles of physics. At least, some theologians maintain that God may do so as and when he chooses, and that this is what is meant by a 'miracle'. But we need not assume that life is a miracle, or not at least in that very simple sense. Life doesn't defy the principle of entropy. Instead, as we saw in Chapter 3, living systems borrow energy from their surroundings. So, while it is growing, the living system reduces its own entropy. But it increases the entropy of its surroundings and the net effect, as must always be the case, is an overall rise in entropy (a rise in disorderliness, that is). Life, in short, is an eddy in the general tide of entropy. It spins backwards, against the general flow, but it is powered by the flow nonetheless.

Now to the point: living systems thrive by taking energy from their surroundings. Sometimes they use this energy to create new entities that would not have been there without them – entities that may include coral reefs or beaver dams or temples or gold watches, but also include rare chemicals that non-living systems would not normally produce. Also, as they extract energy from their surroundings, they are destructive; they make a mess, including a chemical mess, again of the kind that would not have happened without them. In general, the changes thus made in the surroundings, both constructive and destructive, are liable to be more conspicuous, and to endure longer, than the living creatures themselves. So perhaps the best way to find life is not to look for living creatures, but to look for the kinds of chemical and physical changes in rocks and atmosphere and in free-standing water (if there is any) of the kind that living systems make, and would not have occurred in their absence. A hunter would call this 'spoor'.

But then the question arises – how do we know what kinds of changes in the chemistry of the atmosphere, rocks, and water are in fact caused by living organisms? How do we know that such changes would not have occurred in any case? To make such a judgment we need a baseline: we need to know what the atmosphere and the rest would be like if subjected only to the laws of physics and the grand principle of entropy. Basic chemistry tells us much of what we need to know – a simple theoretical and empirical knowledge of how chemical elements and molecules do, in fact, behave when left to themselves. Other kinds of science, including the Earth sciences, round out the picture.

So we can say with absolute certainly that the atmosphere of the Earth is so peculiar in its composition, that it could not possibly maintain its present state without a huge and constant input from living systems. There is, says Lovelock, 'a persistent state of disequilibrium'. Indeed, we can see the present atmosphere virtually as a product of life; and since the atmosphere interacts so freely with the oceans and with rock, we can see the whole outer surface of the Earth as a product of life – or indeed as an extension of life; a grand example what Richard Dawkins has called 'the extended phenotype'. The atmosphere of Mars, though, alas, is just the way we'd expect it to be if there was no life, nor ever had been. It's mainly carbon dioxide, which in the absence of anything else to make use of it, is the end of the entropic line. 'There is', says Lovelock, 'no sign of exotic chemistry'; although 'the implication that Mars was probably a lifeless planet was unwelcome news to our sponsors in space research'. But never mind. There are billions of other stars out there in our own and billions of other galaxies which between them harbour billions of other planets. There must be life somewhere. For as the late French-American microbiologist René Dubos has commented, 'Life is a property of the universe'.

But let us look briefly at the kinds of changes we see on Earth that show the influence of life. They are impressive enough. Yet even more impressive is the climactic thesis of the Gaia hypothesis – *that the kinds of changes made by life on Earth are of the kind that enable life to persist*! In other words, Earthly life as a whole – Gaia – is homeostatic. This insight is more than impressive. It is transformative. There was biology before Gaia, and biology after Gaia, just as there was biology before and after Darwin. Gaia, like Darwin's idea of evolution by natural selection, is a notion that spreads far beyond biology to pervade our entire worldview.

Here we begin to confront the other great no-no of modern science. We are not allowed to think teleologically – as if there was any goal in life, or in the universe as a whole. Our task as scientists is to discern what is the case. We then must seek to work out the underlying mechanisms but we are not allowed to suggest that any of what we observe actually has any purpose, or that it functions according to any pre-laid plan. That takes us deep into the realms of metaphysics, beyond the brief of science. In practice, though, it is hard to avoid teleology. For instance we might observe, as scientists, that haemogloblin (Hb) in the blood combines with oxygen and that it releases that oxygen when it reaches tissues that are deprived of it. So it carries oxygen from the air to the tissues. That is mere mechanistic observation. We may also observe that once in the tissues the oxygen is combined with sugars (by a very complicated route to ensure that the reaction is safe) and so releases energy. But if we say that it's the job of Hb to carry oxygen to the tissues so that they can practise aerobic respiration – that begins to look teleological. It implies that the Hb has a formal job description, written out by some guiding intelligence. So the general advice is, don't go there.

The science of Gaia (even without the metaphysical implications) is immensely complex, and is expanding by the week. James Lovelock's latest edition of *Gaia,* reprinted by Oxford University Press in 2009, is a handy introduction. In *The Animate Earth,* Stephan Harding puts elegant flesh on the bones. The following is just a sampler – though just enough, I hope, to make the point. But as you will see from this brief account, however warily we seek to avoid teleology, it is hard to avoid it. Lovelock himself, in *Gaia,* warns us against it but indulges in it nonetheless (although he assures us that this is just for narrative convenience). So...

'A persistent state of disequilibrium'

Take methane. There isn't much of it in the atmosphere but it's surprising to a chemist nonetheless that there's as much as there is because, in sunlight, methane reacts rapidly with oxygen to produce carbon dioxide and water. Indeed, to maintain the present levels of methane in the atmosphere the world would somehow have to add at least 500 million tonnes each year. But then again, since it takes two molecules of oxygen to oxidise one molecule of methane it would need to add a commensurate amount of oxygen (which in fact by my calculations means around 1000 million tonnes) to maintain the oxygen. Since the concentration of methane and oxygen remains constant year by year this means that such quantities *are* being introduced. Could this be achieved by non-biological – 'abiological' – means? Emphatically not, says Lovelock. Basic chemistry tells us that this is 'improbable ... by at least 100 orders of magnitude'. This means there is only one chance in 10^{100} – 10 with 100 noughts – of this being achieved without living systems to help things along; and 10 with 100 noughts is a ludicrously large figure. Only living systems, borrowing energy from their surroundings for their own purposes, could pull such a trick. So the year-on-year constancy of methane and oxygen, one at a low level and one at a high level, means that there must be life. The same is true of the trace gases that are compounds of nitrogen – nitrous oxide and ammonia. They too must be produced in vast quantities even to maintain the present, minute concentrations, for they react and turn into something else almost as soon as they appear. Again, only living systems could produce the necessary amounts.

Intriguingly, too – for the time being, I put it no higher than that – almost all of the gases in the atmosphere in some way help to keep the Earth as a whole in a state that Earthly life finds agreeable. Oxygen, roughly one fifth of the atmosphere, enables Earth's creatures – or those at least that are adapted to it – to respire very efficiently. That is, when properly controlled, oxygen almost literally 'burns' energy-rich organic molecules including sugars to release the maximum amount of energy of which they are capable in the shortest time. Creatures that can use oxygen in this way – the 'aerobic' types – are thus by far the most active. They can't live in airless swamps as the anaerobic types can but they have many other advantages. Oxygen, though, is if anything too lively. Aerobically respiring creatures need to maintain a whole battery of anti-oxidants to mop up any oxygen that leaks around the body, otherwise it is highly destructive. Several of the vitamins, including vitamin C, are primarily anti-oxidants. Excess oxygen in the world at large is just as dangerous. As things are, forest fires and bush fires worldwide are common – indeed they are key components of grassland ecosystems. They are triggered in the wild by lightning. But every one per cent rise in the concentration of atmospheric oxygen increases the chances of such fires by 70 cent, and if the concentration of oxygen in the atmosphere rose to 25 per cent – not much higher than it is now – then everything flammable would forever be catching fire. At least on land, life would be very difficult indeed.

Nitrogen is equally critical in a different way. Since it accounts for four fifths of the atmosphere and is *relatively* inert we might suppose that once molecular nitrogen gas has formed, by whatever means, then it will simply stay as nitrogen gas and so will simply accumulate; as if it was like argon, only commoner. But that is far from the case. Nitrogen is less active than oxygen but it is certainly not totally inert and if it was left entirely to itself then most of it would combine with oxygen to form nitrate, and dissolve in the sea. The huge amounts of nitrogen gas in the atmosphere are another product of Gaia. Nitrogen gas is produced mainly by 'denitrifying' bacteria in soil and sea which remove the oxygen from nitrate. The vast quantity of nitrogen gas in the air maintains a high atmospheric pressure – which apparently helps to stabilise the climate. It also dilutes the oxygen – nature's 'fire extinguisher', as Lovelock puts it. Chemically speaking, too, adding more nitrate to the sea would be like making it more salty. It would increase the sea's osmotic potential – to the point where most organisms of the kind we know now would find it impossible to live. Then again, although nitrate is an essential nutrient of plants and the plant-like phytoplankton, too much of it is toxic. All in all, from Gaia's point of view, it is much better to store nitrogen in the form of the *relatively* unreactive nitrogen gas, in the air, when, just by hanging about, it brings various benefits. Then, when it is needed as a nutrient, lightning and various bacteria, living in the sea or the soil or in the roots of legumes and alders and other such plants, will 'fix' it to make ammonium and nitrate. So the nitrogen in the atmosphere serves as a superabundant food store – trillions and trillions of tonnes of the stuff that's needed to make proteins and nucleic acids, held in a state where it won't decay, added to and reduced by trillions upon trillions of bacteria of many kinds in soil and sea which between them keep everything on an even keel.

Carbon dioxide provides plants with the carbon that is the prime element of all organic molecules, which in turn serve both as sources of energy and to make the fabric of living organisms – plant cell walls, animal muscle, brain, and all the rest.

As everyone knows these days, too, carbon dioxide is in practice the world's principal 'greenhouse gas': like the glass in a greenhouse, it inhibits the escape of infra-red radiation from the ground. The more there is of it, the more the world warms up. Right now, and for the past couple of centuries and especially in the past few decades, the concentration has been rising rapidly. It wasn't much more than 300 ppm when I first started chemistry in the 1950s but by April 2012 we were up to 391 ppm. Climate scientists tell us that if it reaches 450 ppm, then the climate could spin into a state of flux in which we could be sure of nothing except that the ecological impact will be huge. Forest could become desert, desert could become swamp, and entire countries could disappear beneath the rising seas, although Britain paradoxically could freeze (as the Gulf Stream ceases to flow). On the other hand, if carbon dioxide was too low, the Earth would freeze; the twenty or so Ice Ages of the Pleistocene coincided with intervals of low atmospheric CO_2 (as we can tell from ancient bubbles of air trapped in the ice of Greenland and Antarctica). If there was no carbon dioxide at all in the atmosphere we'd be as cold as the moon. As things are (or would be, if we weren't messing things up) the concentration is just about right.

Ammonia is vital too, however unlikely this may seem. It is present in such low quantities only because it is so reactive and so is soluble in water. Because ammonia is so reactive its role is crucial. As a highly soluble, alkaline compound, it helps to raise the pH of the sea to around 8, which living creatures prefer. Without ammonia, pH would fall to around 3 – roughly that of vinegar. Some creatures could live in this and many more would adapt but nothing could possibly maintain any kind of calcareous skeleton: no shells, no bones, no teeth. Here again, it is hard to avoid teleology. For in order to maintain the required throughput of ammonia, Gaia must generate about 1000 million tonnes of it each year. It all comes from biological sources – excreted by living organisms, or from the decay of dead organisms, primarily by the breakdown of proteins. But those proteins have to be created in the first place – which expends an enormous input of energy. All this seems like a terrible waste but we see, when we look at the system as a whole, that it's worth it. The energy invested in making protein and then apparently squandering that protein pays off, because it makes all life possible. This account of things is shockingly teleological but it describes the facts of the case nonetheless. There is no a priori reason for rejecting such an account except that we have been told to reject teleology. We will look at this again later.

Several other trace gases in the atmosphere (the kind we didn't mention in elementary chemistry) turn out to be conveyors of essential nutrients from one ecosystem to another. Notably, rainwater is constantly washing essential elements from land to sea, including sulphur (a small but vital component of complex proteins) and iodine (the key ingredient of thyroid hormone). They find their way back to the land, it seems, largely in the form of dimethyl sulphide and methyl iodide, both produced by seaweeds and other marine algae. It's they (not ozone, as folklore has it) that provide the fine tangy scent of the seaside.

So terrestrial life depends to a significant extent on seaweeds. Who'd have thought it? I'm a great fan of organic farming but I have often thought that the recommendation to fertilise the soil with seaweeds, as found in many a book of gardening, is a little cavalier. What of the marine and shoreline ecosystems? They and the creatures they contain are important in their own right – not to be pillaged willy-nilly. Since it now transpires that seaweeds replenish the terrestrial world with essential minerals there's a selfish reason for concern as well. Methyl chloride also rises up from the sea and helps to regulate the ozone layer in the stratosphere. The ozone layer protects the world from cosmic radiation and we don't want it to get too thin. But it shouldn't be too thick, either. A little cosmic radiation, like a little of most things, does us good. Here is another life-dependent mechanism that keeps the world fit for life. Gaia in action yet again.

It is all very convenient. The whole system is immensely complex but if any one component was slightly different, it seems that the whole thing would collapse. Physicists say the same thing about the various physical constants that keep the universe intact – the magnitude of gravity, or the weak and strong nuclear forces, or the cosmological constant, and so on and so on. Physicists call this 'the Goldilocks principle': not too little and not too much. Just right. Physicists have produced no plausible explanation for Goldilocks, try as they might. The neatness of the universe, both living and non-living, remains mysterious. As Ludwig Wittgenstein commented, the perennial question remains, 'How come?' Gaia is the biological equivalent of Goldilocks.

The one seriously vital element that is not returned to the land in a conveniently gaseous form is phosphorus, which most commonly exists as phosphate, and is an essential component of DNA and RNA. The general depletion of easily mined phosphate rock is a growing concern for agriculture worldwide (far more important in the grand scheme of things than oil). These days phosphorus is lost even more rapidly than in the past as human beings flush their own phosphorus-rich ordure, and their phosphorus-laden soap powders, into the oceans. There it is consumed by marine organisms; none of which produce any P-rich gases.

Even so, there are mechanisms for returning it. For sea-birds eat the P-rich fish and deposit P-rich poo on land or at least on offshore islands in the form of guano (which was once the source of a financial bonanza). Salmon and trout and a few other fish also bring phosphorus back to land by migrating up rivers to spawn. This may sound trivial: but the great coniferous forests of America's north-west Pacific coast evidently get a third of their phosphorus from salmon that come to spawn and are then slaughtered by bears and wolves who then scatter their phosphorus-rich dung through the woods.

All the elements that wash away from the land help to make the sea salty; and more salt wells up from the cracks between the expanding tectonic plates in the ocean bed. The overall concentration of salts in seawater is around 3.4 per cent – which is an awful lot, given that the oceans cover three quarters of the Earth at an average depth of 3.7 kilometres. Ninety per cent of the salt is the familiar sodium chloride, the stuff of salt cellars, while the other 10 per cent is mostly magnesium and sulphate, with some calcium and bicarbonate. In practice, the sea doesn't contain discrete entities of sodium chloride and discrete entities of magnesium sulphate or calcium bicarbonate. All these salts are in dissociated, ionic form – the three metals (sodium, magnesium, and calcium) as positively-charged ions; the three non-metal radicals (chloride, sulphate, and bicarbonate) as negatively charged ions. Silicon, whose qualities are both metallic and non-metallic, is also a key player. Other ions, including the non-metallic bicarbonate, and the hydrogen and hydroxide ions that water itself provides as it partially dissociates, complete the mix, and ensure that the electrical charges balance out.

Old-style textbooks told us that the sea is getting saltier and saltier, as the rain continues to fall and carries more and more salts into it. Since we know how rapidly this occurs – about 540 million tonnes are added each year – we can calculate the age of oceans from how salty they are. And the answer is, when we do the calculation – eighty million years! If we add in the salt that wells up from the sea-bed, it comes down to sixty million years.

But of course there's a snag. These figures are obvious nonsense. All other evidence suggests that the oceans are at least 3.5 billion – 3500 million – years old, which is when life began (for it probably began in the sea). Indeed, says Lovelock, a calculated age of sixty million years is not too far removed, conceptually, from

Archbishop James Ussher's suggestion, from the 1650s, which he based on the chronology of the Bible: that the Earth had been created just before October 23, 4004 BC. In both cases – back to the drawing board.

It's clear, then, that most of the salt that is added to the sea in such prodigious quantities must in some way be precipitated out, so it doesn't continue simply to make the sea saltier. In other words, there must be a 'sink'. So indeed there is and, as ever, the necessary mechanism is provided by living creatures. In particular, a range of single-celled protists (aka 'protozoa') in the plankton, including foramens and radiolarians, extract calcium and silicon from the sea to make their skeletons. When they die, their calcium- and silicon-rich skeletons sink to the bottom to create the chalk, limestone and flints that make up so much of the modern continents. This wholesale removal of calcium and silicon from the sea has a chemical knock-on effect – for without calcium and silicon to maintain the overall electrical balance, sodium and chlorine can no longer be held in solution in ionic form. In the still warm waters at the ocean's edge the sea evaporates, and its salts become more concentrated, and finally the sodium and chloride ions, already in a state of electrical tension, are forced out of solution, bound together as sodium chloride; and so we find great salt-beds all around the world, the spoor of seas long past. (This, at least, is a crude précis of the far longer and intricate account that Lovelock provides in *Gaia.* Chemistry in real life gets very complicated.)

Other evidence suggests that in reality the sea has *not* got saltier as the aeons have passed. Thanks to the various planktonic protists, it has probably remained at about the same level of saltiness for the past several billion years. This is just as well, for as Lovelock says, if the salinity reaches six per cent then most creatures die: only a few specialist microbes and super-adapters like the brine shrimp can withstand the osmotic 'pressure', which would suck most creatures dry. Too much salt, too, stops metabolism; the complex organic molecules cannot properly interact when they are crowded out by too many ions. Once again, we see Gaia at work.

So what's going on?

What do we conclude from all this?

First, that life and the Earth are ultimately mysterious. Some scientists clearly think that if they keep at it then eventually we'll know all there is to know. Certainly, present-day knowledge is prodigious. But the chief lesson from the past few hundred years of science is the one that Socrates drew from his decades of philosophy: that the more we know, the more we see there is to know. Logically, too, we cannot know how much we don't know, unless we are already omniscient, and can compare what we do know with what there is to know. All humanity, including scientists, has to learn to live with mystery; and with mystery should come humility. (Shouldn't it?)

As scientists, we need to ask how the overall facts of Gaia square with known science. At a purely mechanistic level we can invoke the general notion of the 'negative feedback loop': as any one activity increases, or any one component of a complex system rises in concentration, so it brings about its own curtailment. The negative feedback loop is a prime device in the metabolism of all living creatures. In our own selves, for instance, as the output of any one hormone such as insulin or thyroxin increases, it triggers the release of another hormone which puts a stop to it (and often we find that the suppressor in turn triggers another hormone which puts a stop to *it,* and so on and so on. The tuning tends to be very fine). In Gaia, in a similar way, we see that as atmospheric carbon dioxide increases then the temperature rises and the rising temperature encourages photosynthesis, which removes it. (But this can work only up to a point. For if the carbon dioxide rises but there is too little water, say, then photosynthesis cannot increase; and if the temperature rises too high then the plant is stressed and photosynthesis is reduced. But the loop of carbon dioxide/ temperature/ photosynthesis is a powerful regulator nonetheless).

In a general, arm-waving way, we could say that the control of the sea's salinity is another negative feedback loop. The rise in silicon and calcium encourages the growth of the protists which remove it – and this in turn regulates the concentration of all the other ions. Yet it all begins to seem too convenient to be true. Why should evolution have produced a suite of creatures that are able to regulate the overall composition of the oceans for the benefit of all? It's a negative feedback loop alright – but how could such a mechanism have evolved?

The point here is *not* simply that the overall system is complex. Many have argued of late that it is impossible to explain the evolution of any complex mechanism in terms of natural selection but as Richard Dawkins explained in *Climbing Mount Improbable*, the explanation is usually easy enough. Thus it has often been suggested that complex eyes like ours could not have evolved step by step because the intermediate stages would be of no use. As the rhetoric has it, 'What use is half an eye?' Yet half an eye is of enormous use, if you have the right half. Retinas – patches of light-sensitive cells – or even single light-sensitive cells are extremely useful on their own, even in the absence of lenses and muscles for focusing. So we find many creatures in nature that have retinas or eye-spots, but no lens. But we don't find creatures with lenses but no retinas.

But in the case of the organism called Gaia, the explanation becomes harder. For natural selection in general favours characters that bring short-term advantage. It favours genes (or individuals or perhaps groups) that by whatever means do things or acquire qualities that enhance their own chances of surviving and replicating. The foramens and radiolarians who acquired the ability to build skeletons from dissolved calcium fit this bill, of course: they are among the most successful organisms of all time – very ancient, and very abundant, as attested by the masses of chalk and limestone that earlier generations have left behind. But it turns out that by looking after themselves these protists make it possible for other creatures to live in the sea – which most, otherwise, could not; and since all the creatures that live on land evolved from creatures that lived in the sea, this means that almost all Earthly life, and certainly all the kinds that we encounter day by day, depends for its existence on foramens and radiolarians. That, at the very least, is a very considerable bonus; a serendipitous side-effect indeed of the protists' attempts to feather their own nests.

At least we can see, though, why foramens and radiolarians should have evolved the means to take up calcium and silicon. Beyond doubt, they benefit. But what do seaweeds gain from their prodigious outpourings of dimethyl sulphide and methyl iodide? What's in it for them? Quite a lot, possibly: though what it is, is not obvious. Isn't it fortunate, too, for the forests of north-west North America, that salmon should bring them such benison of phosphorus? Why don't they spawn at sea, like most fish? Obviously natural selection does favour the salmon (and trout and sturgeon and so on) that prefer to live in the sea yet breed in rivers. But again, isn't it lucky that it does so?

In many cases, indeed, it is tempting to think that natural selection is operating not simply for the benefit of individual creatures or their genes, but for the benefit of Gaia itself. The whole thing works because particular groups of organisms, and all the organisms taken together, behave in ways that benefit the whole. Again, this tempting thought leads us into strange, in some ways uncharted metaphysical territory. So to what extent should such temptation be resisted? And why?

We will look in more detail at these issues in Part II. First, in Chapter 6, I want to ask what all the ideas we have looked at so far throw light on the present state of humanity.

6. The Human Condition

What difference do all these notions make to us? The fact of evolution in general; of nature's essential cooperativeness; of animal intelligence and consciousness; of the essential *niceness* of animals that are both conscious and social?

Some – including many traditional theologians and many who are steeped in what are called 'the arts' or 'the humanities' would say, 'Not much'. Such thinkers, steeped in a Platonic-Judaeo-Christian tradition, take it to be self-evident that human beings are special – that we are not like the rest of nature: that we are separate; that the rules that may or may not apply to other creatures simply have no relevance to us. Many still reject the idea of evolution. They are especially offended by the notion that human beings and chimpanzees share their ancestry – especially as their most recent common ancestor, the metaphorical grandparent of us both, apparently lived no more than seven million years ago; a twinkling of evolutionary time. There are popular songs in the southern United States (quite catchy, some of them) that rubbish the whole idea. As for the notion that properly began in the western tradition with Darwin – that *all* Earthly creatures share a common ancestor: that sea urchins and mushrooms and oak trees are our *literal* cousins and third cousins; that, many feel, is simply absurd.

Many others – including most biologists but also including many theologians and other thinkers – accept the general idea of evolution and happily acknowledge our affinity with apes, and indeed, more distantly, with mushrooms. But they agree with Descartes that thought and subtlety of emotion, consciousness and sensibility, are unique to ourselves. Any other suggestion, they feel, is mere 'anthropomorphism'. So although they do not insist on an absolute biological divide between 'us' (humans) and 'them' (the rest of nature) they nonetheless see a yawning rift of a qualitative kind – not just a matter of degree. We simply should not extrapolate from them to us, they say. Once again, then, what other animals do seems to have very little relevance to ourselves.

Others, including the neo-Darwinians Mark II, now in the form of ultra-Darwinians, take it to be self-evident that our view of life must above all be 'rational', and for various reasons (discussed later) they have it in their heads that rationalists must be pessimists. Perhaps they feel that if you look at things in the worst possible light then you are least likely to be disappointed. In any case, they have built a body of theory that seems to show that life is indeed as Darwin said it is – one long 'struggle'; or as Tennyson said, 'red in tooth and claw' – and that the whole caboodle is driven by the need for 'selfish' DNA to replicate. Competition is the thing, in this ultra-Darwinian vision. Cooperation works only as a temporary device, a fragile tribal alliance, enabling the cooperators to bash some third party before they get back to bashing each other. Generosity is perceived purely as self-interest in disguise. Indeed, the ultra-Darwinian orthodoxy has it, animals are generous only to individuals who share the benefactor's genes and so, in effect, are extensions of the benefactor – or, at best, they do favours only to those who they are sure will return the favour ('tit for tat'). So these neo-Darwinians are happy to accept that we are part of nature – but reject any notion that nature is nice. To be a part of nature, they suggest, is to inherit a propensity for strife and viciousness. Sometimes niceness rises to the surface as a Machiavellian tactic but as Hamlet observed of Claudius, and as Machiavelli recommended, one may smile and smile and be a villain.

Then there's David Hume's point: that whether nature is nasty or nice, it makes no difference to how human beings *ought* to behave, because what nature does has no bearing on what is morally right. In similar vein the early twentieth century Cambridge philosopher G.E. Moore spoke of 'the naturalistic fallacy' – meaning that it is wrong to suppose that we can derive moral guidance from nature. Again, more later.

Then, finally, many simply look at human history and at our present condition – the injustice, the cruelty, the wars, the genocide, and all the rest – and conclude that human beings must indeed be a bad lot. There is no other plausible explanation for all that vileness. Whether or not this has anything to do with our relationship to the rest of nature does not seem to be relevant. Those who take this view also conclude that we are kept from each others' throats only by the strong hand of leadership, in turn informed by scholars of superior wisdom. The world's rulers, politicians, and academics tend to favour that idea. It gives them their excuse for staying in charge and drawing their salaries. They are also at pains to spread this idea – partly, to be fair, because they think it is true, but partly too because it helps to persuade the rest of us that we need their ministrations, and a firm hand. (Relevant here if not precisely to the point, is Berthold Brecht's observation that the rich need the poor more than the poor need the rich, but the poor don't realise that yet).

In short, a great many scientists, theologians, philosophers, and indeed politicians, would doubt if the ideas in the first few chapters of this book are actually true; and some at least of those who did feel that there might be some truth in some of them, would question whether they had any relevance to the human condition. The standard view is that nature is not nice, and we are not nice, and even if nature was nice it wouldn't make any difference to us. But of course I think the ideas as presented so far *are* true. Certainly, as I will argue in Chapter 10, they have at least as strong a claim to truth as the orthodox, materialist, neo-Darwinian view. I also maintain that they are very relevant to the human condition. Thus...

Speech and sociality

It is remarkable to what lengths some biologists as well as non-biologists have gone these past few centuries to conserve the ancient conceit that human beings really are of different clay from all other Earthly creatures. Increasingly, though, it seems that Darwin was right in this as in so much else: that in almost every respect the differences between us and them are only matters of degree. Certainly it is very hard to find anything that we do that some other animal, somewhere, does not do also (although we do have the edge in versatility. We do a whole range of different things *each* of which has some precedent in nature. But no other creature can do as many different things as we can, or switch so readily from one mode to another. We above all are the proteans). Thus for a long time biologists clung to the idea than we human beings are the only tool-makers. So when Professor Phillip Tobias of the University of Witwatersrand named what he took to be the earliest *bona fide* human fossil in the 1960s, he called it *Homo habilis* – handy man. What made it a true *Homo* was that its fossils were accompanied by stone tools, which presumably it had made. But even tool-making does not truly define us. Now it is known that chimpanzees also make tools (from stone as well as from twigs and leaves) and so do Caledonian crows (in the wild from thorns, in the lab from wire) as outlined in Chapter 5.

Yet there is one thing that really does define us – as first formally pinpointed by Descartes. Humans, he said, are different from the rest because we have speech. In this he was surely right – even though he was spectacularly wrong on many other aspects of psychology. Notably he was wrong to suggest that verbal language is essential for thought of any kind, and that because animals cannot speak they therefore cannot think in any worthwhile sense, and (for good measure) that they therefore cannot feel in any recognisably human sense. This latter idea seems intuitively to be wrong and much research has now shown that it *is* wrong. For one thing, human thought itself is not so dependent on words as Descartes supposed. We express our thoughts through words right enough, but the initial formulation of those thoughts must be wordless. We just couldn't think as fast as we obviously do if we had to spell out ideas in words, albeit in words inside our own head. So it was that Barbara McClintock (of whom more in Chapter 10) said that she grasped the notion of jumping genes in a flash – but it took her a couple of hours to explain it to her colleagues, in words and sentences that are ponderous compared to the speed of thought itself.

As outlined in Chapter 4, chimpanzees can learn a lot of human words and parrots can actually speak them – but in neither case is speech their natural medium. These clever creatures are merely adjusting, *ad hoc,* to the eccentricities of their human companions, for want of anything better. In any case, speech is not the only form of language. We, human beings, obviously communicate a great deal of essential information without speech. Most obviously we do this by deliberate signing – we all of us know how to point and beckon and so on and deaf people can communicate rich and complex ideas by signs. We also communicate a great deal about our underlying mood and our *true* meaning (for example, whether we are lying or joking) by subconscious clues, including involuntary facial expressions and 'body language', from hands on hips to twitching feet. We are also more olfactory than we give ourselves credit for – the cosmetics industry seeks to turn human pheromones into big business. In us, such signs and signals play second fiddle to speech (or so at least we think they do). But it's clear that we can also convey a great deal of information without speaking and there is no reason to doubt that animals do too.

According to Noam Chomsky, however, what really makes human language superior is our use of syntax; not just a vocabulary of sounds and signs but a set of rules, built into our brains, that enable us to arrange the words and signs in a potentially infinite range of permutations so as to convey a potentially infinite range of meanings. Other animals can't do this, so some psychologists have argued. They can convey meaning by language right enough – albeit non-verbal language – but only in a limited way: essentially by reading from a set menu of predetermined meanings. They cannot rearrange the menu at will to convey an infinite variety of meanings. In short, the important distinction between the language of humans and the language of other animals is not that our language is based primarily on words, and theirs isn't. It is that our language is underpinned by the rules of syntax which make it infinitely flexible, and theirs isn't. That is: we have syntax and they don't.

I reckon that this ability alone – our ability to express our thoughts precisely and relatively quickly through a word-based language that in principle is infinitely flexible – is what gives us our huge ecological advantage over all other creatures. For this enables us to *share* our thoughts to a unique degree – in principle, with all other human beings. Once writing was invented, and words could be presented in visible and permanent form, we became able in principle to share the thoughts of all other human beings that have ever lived (or at least of those who lived in the age of writing and had mastered its skills). So each of us becomes plugged in to one vast human brain that extends back in time for thousands of years.

Indeed it is not our individual intelligence or our manual ingenuity that gives us the edge over other creatures. If any average city dweller was matched against a hyaena in a task that involved, say, making a living in the jungle while avoiding traps set by wily zoologists, I would back the hyaena every time. In one-to-one contests, even contests of intellect, we would not necessarily win. But, as human beings, we don't have to face any problem on our own. We can read books on jungle lore and indeed on animal traps, and thus enter almost any situation, forearmed and forewarned. More generally, when we're faced with any problem, we can just ask somebody. Other creatures too learn from their elders, of course – young sheep need old ewes to guide them round the hills; herds of elephants rely almost absolutely on the female elders, and so on and so on – but none apart from us has access to the experience and wisdom of all our fellows, past and present.

As primates, we are innately social animals. Sociality really is in our DNA. But we more than social. As Matt Ridley says in *The Origins of Virtue,* we are *eu*social: which in this context means super-social. It not just that we enjoy the company of other people. In practice we cannot live without them. Like those other supremely eusocial creatures – ants, bees, and termites – we practise division of labour: we divide up life's essential tasks between us, each for the most part performing only a part of what's needed to keep ourselves alive. *Between us* we do all that's necessary. Robinson Crusoe's obsession when he was first marooned was to find another human being – which he did in the form of Man Friday. But even before Man Friday turned up he relied absolutely on artifacts from the stranded ship – things made by other people. The same in principle is true of ants: life's essential functions are divided between the reproductive queen and the fertile males, the protective soldiers (in some species there are several classes of soldiers) and the various classes of specialist worker.

But there is an obvious and crucial difference between our eusocial selves and the eusocial insects. Insects are simply programmed to be eusocial. We too have a genetic predisposition to be eusocial but we also have choice. If we choose, we can be unsocial – as some people do. Some elect after all to become hermits (although we all recognise that this is immensely difficult to do – and all hermits that have made it into history have in practice relied quite heavily on the rest of humanity). Nonetheless, the fact that we can in principle choose not to be social means that for us, eusociality is a matter of choice. So in this too we are unique. We are the only creatures on Earth that consciously *choose* to be eusocial.

Then again, although speech is surely not essential for thought, it is surely a significant aid; it helps memory and increases precision. In fact I reckon that speech and intellect must have co-evolved, each feeding on the other – just as it has often been suggested that hand and brain co-evolved in our earliest ancestors. The bigger our brains became, the more we could develop verbal language; and the more we developed the power of words, the more our brains were encouraged to become more complex. This is yet another positive feedback loop of the kind that leads to rapid evolution. In short: speech is not essential for thought but it is what enables us to think as clearly and as broadly as we do, and it enables us to cooperate as much as we do. It enables us to become eusocial by choice – what a biologist might call 'facultative eusociality', as opposed to the 'obligate eusociality' of bees and ants. Common sense and theory based on computer modeling demonstrate that by cooperation any kind of creature, including us, can achieve far more than we can alone. Indeed it is our capacity for cooperativeness and in particular for sharing thoughts, that makes human beings so successful as a species.

In the light of all this it is perverse in the extreme to base our politics and our economy on competition – the converse of the ability that makes us so successful as a species. Doubly perverse is our insistence on intellectual property rights, which can make it impossible for scientists even in the same field to share ideas, on pain of dire punishment. The idea that competition is good and indeed is necessary has grown up for a whole host of reasons, as outlined in Chapters 1 and 2 – but most of those reasons have been primarily political and based on political and philosophical preconception. The idea that out-and-out competition at the expense of cooperation actually benefits humanity and the rest of the world, does not stand up to serious scrutiny. Common sense, common morality, and sound biology (as opposed to the crude, ultra-Darwinian kind) proclaim that we should abandon the western obsession with competition with all possible speed and focus on cooperation. That is the only way forward.

I am writing these particular paragraphs while the London Olympic Games are in full swing. The games, we are often told, are 'all about competing'. Well, up to a point. When the athletes are interviewed after their events they always say how wonderful it is to win, of course – but they also stress that what really matters is the camaraderie; the sense of belonging to the great global family of athletes, and how much they owe to the crowds who cheer them on. They all stress, too, the cooperation that is needed between athletes and their coaches and families, and the organisers and the volunteers, to make the whole thing work. The competition, in the end, just lends spice to the party, just as an edge of polemic makes any conversation more interesting. The corporates that cash in on such events, pushing out all commercial competitors with exclusive deals on catering and the rest, strike a false note – as is evident to everyone except those politicians who themselves depend on corporate support, and are intellectually committed to the super-competitive modern economy. Thus, human nature and history are misconstrued and misrepresented even as they unfold before our eyes.

Most people are nice

We need not and surely should not regard science as the ultimate arbiter in our search for truth (see Chapter 9) but it is certainly good to have science on our side; and there's more and more evidence (and a growing pile of learned treatises) to suggest that people are, after all, basically cooperative, and basically nice, and that this is how they prefer to be. The latest at the time of writing is in *Nature*.[22] In a series of 'economic games' David G. Rand from Harvard and his colleagues gave each of a group of subjects a pile of hypothetical resources and then asked them whether they would prefer to keep all the goods for themselves, or put at least some into a pool to be shared among everybody. On average, the subjects elected to donate at least half of their booty to the common pool. But also – and here's the really intriguing bit – some of the subjects were told they must decide very quickly, while others were given more time. The quick-responders donated far more of their hypothetical wealth to the common pool – in fact they averaged 67%. The ones given more time donated on average only 53%.

The explanation – based not least on the thesis expounded by the Israeli-American psychologist David Kahneman in his *Thinking, Fast and Slow* seems to be that when people are obliged to make up their minds quickly, they rely on intuition; but when they have time, they engage their conscious reasoning.[23] The results suggest (though these are my words, not David Rand's) that our intuitive selves are much nicer, altogether more concerned with the general good, than our thinking, conscious selves; and indeed – given that out intuitive selves in these tests gave two thirds of their worldly goods to the common good – are eminently agreeable. But our thinking selves get in the way. This seems to be to be the precise opposite of the Enlightenment view – which still seems to guide modern politics – which says that human beings are basically nasty until the smoothing hand of reason is applied.

David Kahneman, incidentally, was awarded a Nobel Prize in economics in 2002 for his ideas on human cooperativeness. It's remarkable how many economics Nobel Laureates there are who have said such sensible and agreeable things – and how the sensible ones have been so assiduously ignored. The one economics Nobel Laureate who seems to be taken seriously is Milton Friedman, the champion of neo-Darwinian neoliberalism. Hmm.

But then, Darwin's ideas when interpreted more sympathetically tell us that natural selection should favour cooperativeness and sociality because sociality clearly aids survival; and so, in practice, it does. All animals are social up to a point and some (including us) are so social that they cannot live without each other. It is obvious that sociality requires give-and-take, so we would expect animals that habitually live in societies to evolve codes of behaviour – what might be called 'manners' – to ensure that each treats the others with proper respect. Again, natural selection has done what we might expect of it. Social animals maintain their societies through all kinds of protocols and rituals. Since sociality requires giving as well as taking, we would also expect natural selection to favour generosity.

Since all creatures prefer to do what brings them pleasure (even the behaviourists agree with this, although they question whether pleasure is 'real') we would expect natural selection to produce creatures that positively enjoy the kinds of behaviour that make sociality possible. We would expect social animals positively to enjoy the protocols and manners, and to enjoy the giving as well as the taking. We would indeed expect natural selection to favour emotions such as affection which ensure that in helping others, even (or especially) at cost to ourselves, there is no hardship at all. We would expect evolution to favour love, in short, to provide the essential social cement. We know that human beings are capable of love – and many have argued from many points of view from the spiritual to the carnal that life is pointless without it. Certainly a loveless life is bleak. Since there is no a priori reason to reject anthropomorphism out of hand, and plenty of reasons not to, it is eminently reasonable to suggest that other animals know how to love, too; and as Frans de Waal has pointed out, the affection between individual creatures is not confined to blood relatives. Indeed, since sexual reproduction in intelligent creatures clearly involves more than simple attraction – Lorenz the great observer speaks of geese and jackdaws 'falling in love' – and since a great many animals clearly go to great lengths to avoid incest, it is obvious that individuals *must* experience affection for non-relatives if they are to reproduce at all.

Where does morality come in to all this? As suggested earlier, and as argued again later, we cannot properly say that an in-built tendency to behave generously and indeed altruistically is *in itself* moral. It seems to be embedded in the idea of morality that it must involve choice. A creature that behaves generously simply because it is 'hard-wired' to do so, or merely gains pleasure from doing so, cannot properly be said to be 'moral'. The truly moral creature must be able at least in principle to consider the possibility that it could behave meanly and that it might gain in the short term if it did so; to consider all this *and yet* choose to behave generously. We can say, though, that an inbuilt propensity to cooperate, and an inbuilt predisposition to be generous, are the foundations of morality. Creatures and animals that habitually behave socially and generously we tend to say are *nice.* Most of us most of the time don't stop to ask whether the niceness can truly be considered to meet the criteria imposed by moral philosophers. For day to day purposes, nice is nice.

In short: good, social behaviour becomes moral behaviour when we are able at least in principle to consider the possibility of *not* behaving sociably and generously, and choose to be social and generous in any case. We know that human beings are capable of this. The people we honour above all others are the ones who devote themselves to the welfare of others, including saints and schoolteachers and doctors and nurses, and war-heroes and firefighters. We should not assume as so many generations of past thinkers did assume, that other animals are not capable of such deliberate self-sacrifice. Animals clearly do behave generously and altruistically, and some of them at least, perhaps, may be perfectly capable of weighing the odds. A rapidly growing body of evidence suggests above all that we should not underestimate our fellow creatures. They have been underestimated in the past not for reasons that stand up to serious scrutiny, but for reasons of dogma and expediency.

The conclusion of all this is that we would expect, on simple evolutionary grounds, that most human beings would be nice. Human history seems to suggest that this is not so: all those wars and murders and betrayals and so on and so on. But, I suggest, the day-to-day experience of most people under most circumstances most of the time supports the notion that most people are indeed nice. I am sure that Anne Frank was absolutely right to observe, while the Nazi occupiers dragged her friends and relatives off to the camps and she and her family were imprisoned in an attic:

> It's really a wonder that I haven't dropped my ideals,
> because they seem so absurd and impossible to carry out.
> Yet I keep them, because in spite of everything I still
> believe people are really good at heart.[24]

Incidentally, Anne Frank was just fifteen when she wrote the above.

Nowadays people travel a great deal – and most people who are not themselves idiots remark when they return from foreign parts, however foreign those foreign parts may be, how *nice* the local people are. In general it's only in societies that for one reason or another are seriously fraught – typically suffering from some invader or some ghastly despot – that people sometimes behave ungraciously. Niceness, for most human beings, is the preferred, and the default position; what most people are usually like except when stressed. Darwinian theory properly applied suggests that this ought to be so; and the common personal experience of most people most of the time confirms that it is so.

Neither is it true, as conventional economists and politicians seem to find it convenient to suppose, that human beings are particularly materialist. Most of us are far less concerned about the size of our car and of our house and garden than we are about our family and friends. Those who put material things first we think are distinctly odd, and indeed are sociopaths. Many a survey confirms all this: that most people are not obsessed with material possession. More sophisticated studies, as outlined in Geoffrey Miller's *Spent,* show that when people do take their cars and their clothes seriously, it is usually for social reasons.[25] They want indeed to indicate that they have some status – and, says Miller, we need not assume that status implies a desire to push the rest of us around. High status, he suggests, comes from being a good chap – like a successful alpha woolly monkey; so a big car is primarily intended to say, 'I am a nice reliable person. That's how I got to be rich'.

Then again, it is true and obvious that if we are starving in the gutter then we feel much better if our material status improves, so that we can eat well and have a roof over our heads. But once we have achieved life's basics, plus a bit more, further wealth does not make us happier (unless we are sociopaths obsessed with wealth). Personal experience tells us this; so does many a folk-tale (as in 'poor little rich girl'); and so again does many a modern survey. It is perverse therefore for modern governments to gear the endeavours of entire nations to the accumulation of wealth. It would not be sensible even if the accumulated wealth was firmly distributed, which of course it is not. Governments pursue this obsessive goal only because they have locked themselves, and therefore locked the rest of us, into a particular form of capitalism that relies on constantly increasing wealth. There is no such thing in the modern economy as a steady state, however agreeable that steady state may be. So we are all committed to make more and more – and to do so in head-to-head competition with everybody else. Demonstrably, this is destructive and inefficient, and it is psychologically perverse. But it is the official line, espoused in high places everywhere, and supported by battalions of highly paid intellectuals. Clearly, those intellectuals believe that an economy based on material acquisition and competition is a good and necessary thing and is in line with human needs and wants. But as George Orwell said in a slightly different context:

> One has to belong to the intelligentsia to believe things like that: no ordinary man could be such a fool.[26]

'Intellectual', alas, does not mean 'sage'.

In short, human beings emerge rather well from even a cursory analysis. Most of us are not particularly materialistic (once life's barest necessities are satisfied, with a bit more for leeway) – though most of us certainly have a sense of justice, and want fair shares; and most of are nice. Why, then, did the thinkers of the Enlightenment generally suggest the opposite? Why did they typically refer to people at large as 'the mob'? Why has Christianity, the world's most widespread religion, burdened humanity with the concept of 'original sin'? Why does human history seem so foul?

Game theory: why nice people are ruled by nasty people

If everyone – or at least, most people – is as nice as I suggest then the world should be all sweetness and light, should it not? Indeed it should be, and could be. We, people at large, Ordinary Joes, have the power to make it so. This is one of the principal take-home messages of the New Testament (or at least, it is one obvious interpretation). But a few little glitches get in the way.

First, all human beings, like all thinking creatures, have a touch of Jekyll and Hyde. We are all capable of anger and even of violence and indeed of meanness and outright selfishness (in the proper sense of the word – not as in 'selfish gene'). In so far as our psychology is underpinned by our genes, they are a mixed bag: what has been called 'the parliament of genes'. Sometimes we are confronted with anger and even with violence and then, if we are not utterly cowed by it, we tend to return like for like. The phenomenon of road-rage is most revealing. I am sure that soldiers returning from war are as traumatised as they often are partly because of what war has taught them about themselves: that they sometimes find themselves committing acts that in normal times they would find unspeakable, and indeed unthinkable. It must be hard to treat your own children normally when you have just returned from shelling someone else's. Sometimes anger and violence seem the only possible responses to life's pressures. Sometimes they seem to bring short-term reward – they remove immediate danger. Sometimes they simply seem to make life easier – an easy way to clear away obstacles that are not actually dangerous, but are merely irritating. Almost all of us, too, as good primates, have a built-in tendency to follow the acknowledged leader, to go with the social flow. So despairing mothers say of their wayward sons (sons more than daughters), 'He's a good lad at heart, but he's fallen into bad company'. Sometimes the mother's analysis is true.

As with every other human quality, however, some of us have more of one than another. Some people, of the kind that in olden times were called 'choleric', are easily roused to anger. Others are more passive and altogether peaceable. Some are more or less indifferent to worldly goods (so long as they are not actually starving in the gutter) while a few really are made happy by wealth (or so it seems), as the standard economists say is true of all of us (or at least they think they are made happy by wealth, which in practice amounts to the same thing). Some are naturally generous and altruistic while others are reluctant to share and always have an eye open for the main chance. Some people are obviously empathic: they cannot feel content even when they should be having a good time, if they know that others are suffering (like Dorothea Brooke in *Middlemarch*). Others simply don't give a damn whether other people, or other creatures, are suffering or not (like Rosamond Vincy in *Middlemarch).* A great slice of philosophy and political musing and alas, some neo-Darwinian science, is devoted to justifying the notion that people and other creatures who suffer and fall by the wayside have only themselves to blame. They just weren't adequate. They were ousted by superior beings; that is, by the people who have ousted them and are now making the excuses. Too bad. But in the words of an old Pete Seeger song – '... but you can't blame me. No, you can't blame me at all'.

As with every other human quality, too (or indeed, as with any complex quality in any species) we find, when we plot the intensity of the various traits against the frequency with which they occur, that we get a bell-shaped curve. Thus we find that most people are able either to be peaceable, generous, and empathic, or to be aggressive, mean, and entirely solipsistic; and they, the majority, occupy the centre of the bell-shaped curved. Just a few – a very few – are seriously saintly, and are more or less incapable of anger, or of selfishness, and share the hurts of other people and of suffering creatures everywhere. (Jesus was not quite of this kind. As quite a few stories in the New Testament illustrate, he was capable of anger. But he advocated passivity). At the other end of the curve are those – very much a minority – of the kind that are commonly called villains, or evil, or sociopaths, or psychopaths. They have no ambition except to come out at the top of the heap, in wealth and in power. They are indifferent to the suffering of others. They have no empathy. Often they have no fear. Such people make excellent gangsters, as featured in Martin Scorsese's *Goodfellas* or in Mario Puzo's and Francis Ford Coppola's *The Godfather* (although, interestingly, the highly intelligent Don Corleone, the Godfather himself, was not an out-and-out thug. He was kind to some people, he would not trade in drugs, and he would rather have been 'respectable' – which his son and heir, Michael, was more than half way to becoming).

In general, the task for all of us is to strike a balance – between the need to fight our own corner and the need and the duty to serve the society as a whole. Serving the society as a whole is in part an act of unselfishness, of true sociality and altruism; and in part is enlightened self-interest, since the more we strive to improve the society, to make it more convivial, the more pleasant and generally easy it will be to live in it. Politics and the economy ought to be designed to enable us all to strike a sensible balance – essentially as Alexander Dumas put it in *The Three Musketeers*: 'All for one and one for all'. Unfortunately, an agreeable balance is rarely struck, at least in modern societies for more than a few years at a time. In some 'centralist' societies most people's individual interests are sacrificed to the state (which in Stalin's USSR was taken to represent the society, though in truth it did no such thing). In other societies, individual self-centredness triumphs over sociality – as in the present, prevailing, neoliberal economy. In the present economy, as Gordon Gekko said in *Wall Street*, 'Greed is good'. Stalinism and neoliberalism are the two lunatic ends of the same spectrum.

But to return to the argument: if the bell-shaped curve was perfectly symmetrical, then it would be the case that half the population would veer towards generosity and compassion, and the other half would veer towards meanness and selfishness. But I suggest that the curve is not perfectly symmetrical – for *most* people are more inclined to be generous and compassionate; that meanness of cruelty are the minority pursuits. So yes – almost all of us are capable of behaving unsociably from time to time. All of us may sometimes be mean and selfish. But that is not our preferred position, and since it is not, then society as a whole ought to be convivial, should it not? What we actually *do*, day to day, in an organised society like ours, is determined largely by the prevailing political policies and the economic theory. Since most of us are nice most of the time, and certainly prefer to be, politicians and economists merely have to adjust their policies and their theories to the realities of our own in-built psychology; to devise political and economic structures that enable us to behave well, and in general to bring out the best of us. But alas, that is not how life works out – for reasons that are easily explained by a simple version of game theory.

Game theory has ancient roots – some see its origins in the Talmud – but it was expressed in formal mathematical terms in the early twentieth century and commonly attributed in large part to the particular genius of the Hungarian John von Neumann. It does what its name suggests: analyses and expresses in mathematical terms the strategies that win games (or indeed lose them or lead to a draw or produce the best compromise). The US military soon adopted game theory in various forms to help it win battles and wars, and John Maynard Smith (as cited in Chapter 2) began to apply it to evolutionary problems in the early 1970s.

One version of the theory, simple but very cogent, is the game of doves *versus* hawks. The doves in essence are the kind of individuals that I am calling 'nice'. They are always keen to cooperate. They eschew violence. If they are attacked, they do not resist. If the happiness and general wellbeing of each member of a society is quantified – say, given a score on a scale of one to ten – then game theory shows that the total score of happiness in the society (individual score *times* the total number of individuals) is greatest in an all-dove society. The reason is intuitively obvious: doves waste no time and energy in fighting, but devote all their energies to making the society more convivial, and all benefit as a result. The conclusion seems obvious, too. We should strive to create an all-dove society; and since most of us naturally veer towards dovishness, that should not be too difficult.

But there is a snag. The all-dove society is vulnerable and hence, without assistance, it cannot be stable in the long term. For nature also throws up a proportion of people who are more inclined to be hawks. Hawks in game theory are selfish, aggressive, and not afraid to be violent. If they want something, they just take it, shoving the rest aside. Their goal, in so far as they have a defined goal, is to come out on top. In game theory, a hawk in an otherwise all-dove society scores very highly indeed on the wellbeing index, because everything he wants, he gets. Indeed, the wellbeing score of the single hawk in an all-dove society is higher that that of a dove in an all-dove society, because he does not have to waste any time negotiating, and does not make any sacrifices on behalf of the whole society. He merely takes, and the doves give way. Thus the single hawk in an otherwise all-dove society has a higher individual score than the individual doves in an all-dove society, who do pay their dues to the whole society. But the individual doves in a society with a bullying hawk on board are miserable, because they are shoved around. So the total wellbeing score in a society of doves with a hawk on board is less than it would be if there was no hawk.

The theory says that in an all-dove society, a hawk or hawks are more or less bound to appear. The hawk may invade from outside, as in all those Hollywood westerns where peaceable villages of dovish farmers are invaded by bandits – archetypal hawks. Or – given that our genes do have at least some influence on our personalities – the hawk may arise by mutation. Commonly, of course, the hawks are bought off and set themselves up as protectors of the people they originally robbed: hawks hired to fend off other hawks. Sometimes the hawks-cum-protectors are content simply to run a protection racket, like the traditional Mafia (who, at their best, to be fair, really did provide protection). Sometimes they are ennobled, like the Japanese Samurai. Sometimes the hawks simply become the aristocracy, or indeed the government. Either way, hawks rise to the top.

But there is a saving clause. Theory tells us that the hawks must *always* be in the minority. The first hawk to invade or to arise in an otherwise all-dove society does indeed have a whale of a time. He takes what he wants with no opposition. But Darwinian theory tells us that as a successful individual the single hawk will enjoy reproductive success, and so will produce more hawks. All goes well for the hawks – until there are too many of them. Then, when a hawk swaggers in to take what he wants he may find himself confronted, not by compliant doves, but by another hawk. Then all hell breaks loose. One or other of the hawks is liable to be killed, and they both suffer even if they both live. Thus, as the number of hawks increases, the average well-being of each hawk goes down. Soon, a threshold is reached. A plausible version of the hawk-dove game tells us that no society is likely to contain more than fifteen or so per cent of hawks. Much above this, and they start to get in each other's way.

Of course this is simplistic. Psychologically it is not quite right. Notably, as excellently described in *The Spirit Level* by Richard Wilkinson and Kate Pickett, isolated hawks in otherwise dovish societies are not particularly happy, even though they may be very rich.[27] At least, the richest people in societies that otherwise are poor are much less happy than somewhat less rich people living in societies that are more egalitarian. One plausible explanation is that most of us, including most of those who behave hawkishly, do have a sense of empathy, and are made miserable when they see that others all around them are miserable (especially when the general misery is caused in large part by the people who have grown rich). We should stress, too, that there is not an absolute and irreversible biological difference between the hawks and the doves. As discussed above, most of us are capable of behaving hawkishly or dovishly depending on circumstance. So again, we see that the task for governments who genuinely seek to make a better society is to create circumstances that encourage dovishness – and it is a huge indictment of the world's most powerful governments, and indeed the most governments of all kinds, that they choose to do the precise opposite. The present emphasis in Europe and the US (and increasingly elsewhere) on personal acquisition and competition, as opposed to modesty and cooperativeness, is a recipe for disaster, and it really is not surprising that disaster is precisely what we have got.

Simplistic though it is, however, the hawk-dove model seems to have enormous explanatory power. Notably, it seems at a stroke to explain the glaring paradox that is otherwise inexplicable. For if it is the case that hawks must always be in the minority – and a small minority at that – then why should societies as a whole so often behave hawkishly, sometimes seeking literally to take over the world? The answer of course is that although hawks are the minority, they nonetheless call the shots – precisely because they are hawks. Hawks are hell-bent on power, which most of us, being dovish, are not; and people who are hell-bent on a particular goal are far more likely to reach that goal than people whose thoughts are elsewhere. Most of us want simply to get on with our own lives, and really don't *want* power, and all that goes with it. Some people like committees but most, including many of the cleverest and nicest, run a mile to avoid them. So most of us are all too content to leave the business of being in charge to those who want to be in charge. But the people who most want to be in charge are the hawks, and they are hawkish.

As good primates, too, we have a built-in propensity to follow leaders – for most primate groups rely heavily on patriarchs and matriarchs. But, as noted earlier, some primates at least (including woolly monkeys and gorillas) get rid of their leaders as soon as they fail to operate in the best interests of the group, or at least they give any would-be bullies a very bad time. According to John Prebble's *Culloden* (Atheneum, 1962) traditional Scottish Highlanders did the same to their chiefs, if the chiefs fell short of expectations. But in modern societies, even those that call themselves democracies, those in power have layer upon layer of protection. Electing leaders is one thing. Getting shot of them when it all goes wrong is something else entirely.

Sometimes, to be sure, we do see dovish people in positions of power or at least of political influence: Mahatma Gandhi, Nelson Mandela, the Dalai Lama, and Vaclav Havel are recent, obvious examples. But, I suggest, all of those (and I would bet, any other examples you may think of) are or were leaders in societies that were oppressed by some still greater power: India by the British; Mandela by the white supremacists; the Dalai Lama's Tibet by the

Chinese; and the erstwhile Czechoslovakia by the erstwhile USSR. Any hawk who tried to rise to power in any of those subordinate societies would be firmly put down by the bigger hawks of the oppressors. In such circumstances, leaders can become leaders only by being outspoken doves. But if and when the oppressors disappear, normal service tends to be resumed. Usually, sooner or later, hawks take over again.

Still, you may argue, many people in high places are not conspicuously nasty. For my part I was pleasantly surprised on my one trip to the World Economic Conference in Davos a few years ago to find that virtually all the CEOs I met were very nice people, truly anxious to do good. But CEOs, despite their title and their trappings of power, commonly do not make the running in their own companies, any more than presidents necessarily rule the United States. The power is behind the throne: and the drivers in any one company very often are out-and-out hawks, precisely because they are so fixated on personal success. Nice and intelligent people then find themselves driven along by the truly hawkish hawks, and find all kinds of excuses for doing so. I once met a very intelligent American Republican, who, very surprisingly, spoke highly of Donald Rumsfeld. Rumsfeld wasn't the monster he often seemed to be, said his apologist. He merely suffered from excess loyalty. He was employed to serve the duo of Dick Cheney and George W. Bush, and did so with all his energy, because he saw this as his duty. Many a politician says, and does, exactly the same. They put party and personal loyalty above their conscience – and make a virtue of it.

Worst of all is that intellectuals of all kinds also hitch their wagons to the hawkish power-machine – and use their learning and their intelligence both to justify the machine and to make it work. This is true of the law. Much of modern trade law, and the laws of land ownership, and the British laws of libel, favour the rich and powerful. It is true of economics. Modern neoliberalism is not the only possible way to run the world, not by a long chalk. Many traditional business people, both producers and bankers, see neoliberalism as a sad perversion of traditional capitalism. But it makes rich and powerful people more rich, and more powerful, and so there are economists on hand eager to demonstrate what is obviously untrue – that the neoliberal global market, offering freedom to corporates, is good for all of us, and can solve all our problems.

Worst of all, perhaps, is that scientists bend their efforts to the support of the hawkish elite. This is most evident, and most dangerous, in agriculture: the pursuit that directly affects us all and which, in the end, will determine who lives and who dies. The current craze among the powers-that-be is for GM crops: crops shaped by DNA transfer *aka* 'genetic engineering'. In truth GM crops are not necessary. The risks attendant upon them are not worth taking. The whole technology has produced nothing of unequivocal value after thirty years of effort that could not have been produced more cheaply and safely by other means. The whole enterprise is buoyed by hype which to a significant extent is simply mendacious: that GM increases yields; that is essential to feed the world; that to oppose the technology is irresponsible; that science has already proved that there are no dangers. It is all untrue and most of it is stark nonsense. Yet whole battalions of scientists queue up to perpetrate the nonsense. They are not themselves liars, for the most part. Many and perhaps most of them, to be fair, believe the hype. But they are paid by rich biotech companies (or work for ostensibly publicly owned institutions including universities which nowadays depend on commercial funding) and if they didn't work on GM they would have no work at all. In reality, the main point of GM technology (and in the end the *only* point) is to increase the power and wealth of the biotech companies who make them, and of governments such as Britain's that rely on corporate support. Specifically: governments like Britain's tot up the earnings of the corporates that operate within their shores and call it 'gross domestic product' or 'GDP'; and as GDP increases they call it 'economic growth – which, in this materialist, neoliberal age, is the only measure of success that they take seriously. The *means* by which the wealth is produced, and whether the wealth actually brings benefit to the society as a whole, or humanity or the world as a whole, is not considered relevant.

Here we see at work two other general phenomena (which, if we cared to put our minds to it, could also be analysed by game theory). One is simply that of natural selection. The lawyers, economists, and scientists who oppose the current trend remain unemployed, or seriously underemployed, and although some may sneak by in academe and others work for NGOs (non-government organisations) most are apt to fade away (and more and more NGOs these days seem to be taking the corporate shilling). The ones who thrive are the ones who support the status quo.

The other general phenomenon at work is that of the positive feedback loop. Corporates use some of their wealth to employ scientists who will produce the kinds of 'high' technologies (such as biotech) that will make the corporates even richer, so they can then spend even more on science that will produce even more tech that will make them richer still – and so on and so on. Governments such as Britain's are content to sit on the sidelines and prime the pumps: using tax-payers money to pay a minority of scientists to do the fundamental research that the commercial scientists can then make use of. As always, the taxpayers take the risks and the corporates take the profits and the government calls the profits 'GDP' and claims success.

Suspect though all this may seem, it wouldn't be so bad if people at large were in fact benefiting. But clearly we are not. To be specific, agriculture that is based on high technology and run by an elite, primarily for its own benefit, is failing and is bound to fail. Clearly it cannot deliver. Even worse is that vast resources are ploughed into high-tech industrial agriculture at the *expense* of agriculture that really could feed everybody well – and provide hundreds of millions (literally) of good jobs, and do all this without wrecking the rest of the world. Here I want primarily to nail the generalisation: that things have gone horribly wrong because policy in general is *not* designed for the benefit of people and the world at large, but to reinforce the power and wealth of the hawkish elite; and that intellectuals of all kinds have joined the hawkish élite, partly because some of them are hawkish themselves, and partly because it is hard to find a secure job anywhere else. The noose tightens.

But there is worse still. For whoever is in charge of any particular society controls information. Governments and corporates shower us with information which in effect, intentionally or not, is propaganda. Worst of all: governments, at least in Britain, control education: and nowadays for example it is hard to find any course of biology that is *not* geared towards biotech, and the commercial joys that go with it. As discussed in Chapter 10, this shouldn't be what science is for. The overall effect of all this top-down propaganda is to create and reinforce the *Zeitgeist*: a worldview based on an idea of reality that is entirely materialist, and which takes for granted that human beings are in a different league and category from all other creatures, that everything else including other creatures is just a resource, and that resources should be turned into commodities to be sold for cash, but that human beings in general are a bad lot and need to be kept in check by a wise and caring political and intellectual elite. But the political and intellectual elite are the hawks, or people who in various ways convince themselves that it is good to work for the hawks, and hawks work for themselves.

Competition and cooperation: short term and long term

The conflict between competitiveness and cooperativeness is, to a very significant extent, a conflict between short-term and long-term. In a society of doves it pays in the short term to be a hawk – although this is a dangerous strategy which is liable to fail in the long term as more hawks come on the scene. In a society in which everyone is hawkish it pays to be very hawkish indeed and see off the opposition as quickly as possible. The trouble here, for the successful hawk, is that he must always spend time and energy in fighting off the other hawks, which is extremely wasteful;

and the people who do useful things, like growing food and building and looking after people, find it hard to get on with their work because of all the mayhem all around them. So excessively hawkish societies are bad for everyone, including the hawks. In the long term the all-dove, cooperative society is best for all – and is more or less the only kind of society in which the concept of the long term is likely to be relevant. Over-hawkish societies are likely simply to collapse, and there can be no long term.

Though some may find the idea strange, Jesus Christ emerges as a very astute game theorist. He advocated all-out dovishness – most directly as reported in Luke 6: 27–29:

> 'But I say to you who hear: Love your enemies, do good to those who hate you, bless those who curse you, and pray for those who spitefully use you. To him who strikes you on the *one* cheek, offer the other also. And from him who takes away your cloak, do not withhold *your* tunic either.' *(New King James Version)*

In Matthew 18: 21–35 Jesus expands on the turning of cheeks:

> Then Peter came to him and said, 'Lord, how often shall my brother sin against me, and I forgive him? Up to seven times?' Jesus said to him, 'I do not say to you, up to seven times, but up to seventy times seven.'

Many have criticised this advice. They point out among other things that it is impractical. In practice, those who are as dovish as Jesus seems to be advocating would be given a miserable time. Others say that extreme dovishness is irresponsible. If you yourself are content to be beaten up, then fair enough. But of you allow a bully to beat you up, then he will be encouraged to go on and beat more people up. Bullies need to be put a stop to. But Jesus isn't a wimp, as he demonstrates elsewhere. He certainly does leap to the defence of third parties. So St John tells us that when various self-righteous defenders of the law are preparing to stone a woman 'taken in adultery' Jesus put them firmly in their place:

> So when they continued asking him, he lifted up himself, and said unto them, He that is without sin among you, let him cast the first stone. *(King James Version,* 8:7)

To be sure, Jesus does not physically attack the would-be stoners, but he certainly stands up to them. Elsewhere (Matthew 21:12) he does use violence to defend the temple against what he saw as commercial interlopers:

> And Jesus went into the temple of God, and cast out all them that sold and bought in the temple, and overthrew the tables of the moneychangers, and the seats of them that sold doves.

My point is that although Jesus may seem perverse to advocate extreme dovishness – Peter certainly thought so – he was, in fact, advocating the kind of strategy which game theory (and common sense) tells us ought to be the best long-term strategy. Extreme dovishness is difficult, of course – much harder than short-term hawkishness – but it is the course that brings the greatest reward. To be sure, Jesus as represented in the Bible seems generally to suggest that the reward will come in the next world, rather than in this one, which many have seen as a serious drawback. But game theory and common sense suggest that if only we could achieve universal dovishness in our various societies, then the rewards would come in this life too: peace; peace of mind; security; freedom to work at whatever brings fulfilment; and the possibility that our species could enjoy at least another million years on this Earth, in the company of other creatures – and then draw breath and contemplate the following million. The prize is so great, and the consequences of uncurbed hawkishness are so dire, that universal dovishness is surely worth trying to achieve, difficult though it obviously is.

The meek should indeed inherit the Earth, as in Matthew 5:5 – at least in the long term; but only if the not-so-meek, focused on the short term, leave any Earth to inherit.

For it is also obvious that societies of doves must protect themselves. Mavericks, bullies, crooks, self-seekers and general parasites do arise – the mutant or invading hawks – and they do need to be kept in check. This is precisely what societies of animals, and societies of tribes in a state of nature – and indeed living bodies, which are societies of cells -- habitually do. Meerkats that fail to pull their weight in protecting the troop are given a bad time. Many an anthropologist has reported that in traditional tribes from Scotland to Africa to North America, chiefs or warriors or anyone who seeks to rise above their station is rapidly demoted. The human body (in fact all living bodies) have elaborate systems to protect themselves both against invading pathogens, and against other body cells that break the bounds and seek to become cancerous. Cancers are the ones that get away.

We could argue that the main trouble with our own present global society is that we do things entirely the wrong way around. In this ultra-competitive, neoliberal, ultra-Darwinian world, hawks are not kept in their place, or – which would be better – encouraged to use their energies for the general good. They are encouraged instead to fill their boots. The hawks, in fact, are encouraged to take over and run the show. The sociopaths are in charge. No wonder the world is in a mess.

From science to metaphysics

So what's to be done? As suggested in the introduction, to put things right we need to dig deep: and 'we' means all of us, for experience shows that we cannot afford to leave our affairs to intellectuals.

'Digging deep' takes us inescapably into the way of thinking that in the western world has been sadly neglected and all-but forgotten: that of Metaphysics. We need to re-address what

Seyyed Hossein Nasr calls 'the ultimate questions': what is the world really like?; how do we know what is true?; and what is it *right* to do, in our short time on Earth?

All this is the subject of Part II.

PART II

THE GREAT QUESTIONS

7. What Is The Universe Really Like?

The grand agenda of metaphysics is to provide a complete and coherent account of all that is – which entails three main questions: 'What is the universe really like?' 'How do we know what's true?' And, 'What is good? How *ought* we to behave on this Earth?' Then there is also a fourth, meta-question: 'Is there any worthwhile relationship between the facts of the case – what the universe is really like – and the moral question – how *ought* we to behave within it?' Are these two questions connected, or are they quite independent of each other?

Question one, so many suppose, we can simply hand over to science. It is the job of science, after all, to tell us what's out there and how things work. In truth, though, science doesn't actually tell us what the universe is *really* like. It does seem to explain the material universe very convincingly – the things we can see and touch, with or without the aid of extra-sensitive scientific instruments, like the microscope and the Large Hadron Collider. But science doesn't tell us, and is not able to tell us, whether the things we can see and touch are all there is. Is there more going on behind the scenes? Is the material universe simply the surface of things? Many philosophers have thought this. Plato maintained that all we can ever really perceive, directly, is the shadow of reality. More generally, most human beings through most of history have felt in their bones and sometimes fervently believed that there is more to the universe than meets the eye – an idea made manifest in the world's many religions. The general idea that there is more to the universe than meets the eye – and more, therefore, than can be dealt with by science alone – is one that I think can reasonably be called 'transcendence'.

Often the general notion of transcendence transmutes into a specific concept of God, or a family of gods. Buddhists get by without any reference to a specific God, although Buddhism in general certainly takes a transcendent view of the world. They certainly do not take it for granted that what we can see and stub our toes on is all there is.

Just to emphasise: the concept of transcendence is *not* to be equated with the 'God of the gaps' idea. 'God of the gaps' simply says that anything that cannot right now be explained by science (whatever 'explained' is supposed to mean) should be ascribed to God. But this of course means that as science explains more and more (or seems to) so God is squeezed out until eventually the deity will be snuffed out altogether, as suggested in the quote from Peter Atkins. But the idea of transcendence says that there are aspects of the universe – crucial aspects – that cannot be explained by the methods of science, including the question of whether the universe has direction and purpose (what it is *for*) and the meta-question, 'How come?' It does seem, though, that the study of consciousness might provide a meeting ground between the theologians who contemplate transcendence, and scientists in the form of psychologists, physicists, and anthropologists. The study of consciousness might indeed emerge as a study of transcendence, at least peripherally.

Many, though, including numerous but by no means all scientists and some philosophers, are out-and-out materialists. They agree with Lucretius: that the universe and all that is within it is just 'atoms and the void'. To them, the idea of transcendence is a nonsense. Since it is impossible to believe in God without having some concept of transcendence, the out-and-out materialists perforce are atheists. Since science does deal rather effectively with the material universe, the out-and-out materialists are prone to argue that science can indeed tell us all there is to know, or all that is worth knowing, because the material universe is all there is. This is part of what Peter Atkins implied as quoted in Chapter 1:

> There is no reason to suppose that science cannot deal with every aspect of existence ... science has never encountered a barrier.[28]

I suggest that the divide between people with a feeling for transcendence and those who reject all such musing is the most profound of all the divisions within humankind. It has surfaced in many different ways, sometimes interesting and edifying, but often vile. Leaders of religions have often persecuted 'unbelievers', and materialist-atheists, not least in Stalin's USSR and Mao's China, have often persecuted followers of religion; and, still, fundamentalist Christians of the kind known as Creationists, and fundamentalist material-atheists who include some professors of science, are wont to engage in public spats. Most of the time, however, most people do not allow their metaphysical differences to become acrimonious, because most people are more sensible. It's not uncommon in many a society for sceptical Dad to wait patiently in the pub while Catholic Mum goes to mass (or some such similar scenario). The difference between their world-views may indeed be profound but they don't let it interfere with the harmony of their lives. Yet the metaphysical divide is profound nonetheless (assuming that both believe what they profess).

The out-and-out materialist-atheists insist and apparently believe that they are seriously hard-nosed – the exemplars of rationality. They also take it to be self-evident that rationality alone is the source of reliable truth. Anyone who takes seriously the idea of transcendence, the hard-nosed materialists suggest, is simply flakey: a dreamer; potentially dangerous. The claims of religion, they tell us, are just 'dogma'. Belief in religion is just 'blind faith'. The out-and-out materialist-atheists don't seem to realise that their own assertions are no more than that: declarations of personal opinion. There is no proof and can be no proof that the universe is as Lucretius said. The out-and-out materialist-atheists may for whatever reason prefer to believe this – but their belief too is a matter of faith, and their insistence that it is the case is dogma through and through.

But this kind of musing belongs in the next chapter – when we ask the second of the great metaphysical questions, 'How do we know what's true?' Here I want to address the first question – 'What is the universe really like?' The search for an answer must include science – of course. The insights of science are wondrous and vital. I feel we should spend more money on Large Hadron Colliders, rather than less – and a very great deal more on all of the life sciences, and especially taxonomy and ecology. But always I feel we should keep in mind that the insights of science are not sufficient. Often, intriguingly, science itself hints at more behind the scenes. Science has often intimated that there is a meta-agenda in the universe, behind what we can see and touch and measure. Indeed, the founders of modern science in the seventeenth century took this to be self-evident: their reason for doing science at all was to discover this meta-agenda, otherwise known as 'God's purpose', just as it is for any religious pilgrim. But science is not equipped to unravel the meta-agenda. Even more to the point, it cannot tell us whether there really *is* a meta-agenda. It does not and cannot tell us if Lucretius was right or wrong. In fact, science does not, at the most fundamental level, tell us what human beings have wanted to know since our ancestors first acquired their reflective minds: what is the universe really like?

The conflict between those who feel that hard-nosed materialism can and must explain all there is, and those who feel that we really must expand our horizons if we are to come close to a true understanding, becomes particularly evident when we start to consider the phenomenon of consciousness.

What is consciousness?

Consciousness exists: indeed, as physicist Peter Russell points out in *From Science to God,* it is the only phenomenon in the entire universe of which we can be absolutely certain.[29] In so far as we know anything at all, we know that we are conscious. Everything else that we think we know about the universe is made manifest to us through our consciousness. Given that consciousness exists, and that it is our sole link with all the rest of the universe, it surely matters. Since it exists, and science is concerned with what exists, then science ought to be concerned with it – what it is and where it comes from.

Yet scientists of an ultra-materialist kind often seem content simply to explain it away. Enthusiasts for Artificial Intelligence (AI) are wont to suggest that consciousness requires nothing but complex circuitry. Consciousness, they claim, could be simulated if we had enough computers working in parallel, some of which would monitor the activities of others to produce a kind of self-awareness. For consciousness *is* self-awareness, they say: not just thinking, but knowing that thinking is going on.

Well, the AI enthusiasts may be right, but they not be. Clearly, a computer that convincingly simulates human consciousness hasn't been built yet, and the idea that this is possible is again a declaration of faith. But the progress of computer science and technology these past few decades has been nothing less than magical. Some years ago a computer won a game of chess against the world's greatest ever player, Gary Kasparov (although Kasparov claims that the computer's handlers cheated by helping the computer out behind the scenes, and he certainly seems to have a point). The bigger test would be the one suggested well over half a century ago by the principal creator of the modern computer, Alan Turing: to create a computer that could hold a true conversation, such that we couldn't tell whether we were talking to a machine or real person. And yet: even if computers do pass the Turing test one day we must always remember, as philosopher John Searle pointed out (albeit with an example that doesn't actually work) that simulation doesn't mean replication. My son does a brilliant imitation of Jimmy Stewart ('Now wait a darn minute there!') but he does not thereby become Jimmy Stewart. The real thing and the imitations of it remain qualitatively distinct.

Other scientists and some philosophers simply use words like 'epiphenomenon': things that happen on the side when something else happens. In effect, some neurophysiologists, and philosopher Dan Dennett in *Consciousness Explained* (Back Bay Books, 1992) imply if not baldly declare that consciousness is the noise that neurons make as they go about their work. Well, that's another barrier leapt!

But a succession of philosophers at least since Plato have argued that consciousness is indeed real, and may indeed be the most real thing in the universe – the source of everything else. What we call reality is just shadows on the wall of the cave, said Plato. All is mind, said Bishop Berkeley in the eighteenth century; matter is our mind's invention. This has seemed to some – especially the ultra-materialist scientists – to be mere whimsy. But science itself began to reinforce this strangest of ideas in the early twentieth century, when the new breed of particle physicists showed that the behaviour of fundamental particles, such as photons and electrons, is profoundly influenced by the presence of the conscious observer. Fundamental particles, it was quickly shown, exist in two forms: as waves and as particles. As particles they occupy a distinct location in space. As waves they very definitely do not. The notion has emerged – not as whimsy but as the inescapable conclusion of a host of experiments – that matter exists in wave-form until and unless a conscious mind engages with it, whereupon the wave 'collapses' to form a particle, a thing in time and space, which the observer observes, and calls reality. What we see of the universe, then, and take to be the facts of the case, is in reality a series of snapshots which we have ourselves helped to create.

René Descartes in the seventeenth century saw how important the concept of consciousness is – as encapsulated in his famous 'I think, therefore I am'. But he went on to propose that the universe as a whole is divided into two separate components – mind, and matter. This kind of idea – of two things working side by side – is an example of 'dualism'. In our own time, Amit Goswami, theoretical physicist at the University of Oregon, suggests that this dualistic way of looking at the universe is quite wrong; and that it has led the western world astray ever since, to our great detriment. Instead, Dr Goswami suggests the idea of 'monistic idealism' – very much in line with Plato, and in principle similar to that of Bishop Berkeley. 'Monistic' of course implies that there is just one primary force at work in the universe, not two; and 'idealism' is a reference to Plato's notion that the universe is basically an 'idea' – a manifestation of consciousness. Thus in *The Self-Aware Universe* Dr Goswami writes:

> ... instead of positing that everything (including consciousness) is made of matter, this philosophy posits that everything (including matter) exists in and is manipulated from consciousness.
>
> The philosophy does not say that matter is unreal, but that the reality of matter is secondary to that of consciousness, which itself is the ground of all being – including matter.[30]

For what it's worth, my own modest contribution to this discussion is that the basic stuff of the universe should be seen as 'mind-matter' – with matter and consciousness envisaged not as separate entities as Descartes did but as two sides of the same coin. There's an analogy here with Einstein's space-time. But I am not a physicist, and fear to rush in. The idea of monistic idealism, as Dr Goswami has presented it and without further adornment, already makes perfect sense of a great deal that otherwise seems to make no sense at all. For quantum physics, from the outset, produced paradox after paradox – or at least they appear paradoxical to us. These include the facts – now undeniable facts – that a quantum object (like an electron or a photon) may appear to be in two places at once; or that quantum objects 'jump' from one place to another with no intervening trajectory – now it's here and now it's there with nothing in between; or that two different quantum objects (like two electrons) can influence each other at a distance, instantaneously – so that there is no time for any message to pass between them. But, says Dr Goswami:

> The philosophy of monistic idealism provides a paradox-free interpretation of quantum physics that is logical, coherent, and satisfying.[31]

For good measure, the logical, coherent, and satisfying vision of the consciousness-led universe chimes most intriguingly with what many philosophers and theologians have been proposing at least since the beginning of written history:

> This reformulated picture of the brain-mind enables us to understand our whole self entirely in harmony with what the great spiritual traditions have maintained for millennia. *(Ibid.)*

Thus, traditional physics and common sense suggest that observation is a fairly passive process: that our brains simply register signals from the outside world like a camera registering light. But quantum physics tells a totally different story. Since we affect what we observe, the act of observation becomes more like a pact between we who do the seeing, and the thing we are looking at. In the same way, philosopher Peter Coates points out in *Ibn 'Arabi and Modern Thought,* that the great twelfth to thirteenth century Andalusian Sufi mystic Muhyiddin Ibn 'Arabi saw observation as an 'encounter': in effect, the active mind in dialogue with what's out there. I wonder too if the opening line of the Gospel according to St John – 'In the beginning was the word' – is intended to convey the general idea that consciousness is primary ('word' being the English rendering of the original Greek *logos*, as in 'logic'). There are many other examples of the same general notion in many other traditions.

This idea – the primacy of consciousness, and the all-pervasiveness of consciousness – also chimes very well with the idea mooted briefly in Chapter 4: that mind should be seen as a property of the universe, and that we, and other creatures, do not generate consciousness within our own heads, but rather partake of what is all around us. Among other things this idea helps to explain how it is that many different, quite unrelated lineages of creatures have *independently* evolved intelligence (which is not the same as consciousness, but is a related concept). Intelligence is a tremendously complex quality, and it seems strange indeed that so many lineages should have come up with it – and in many different ways, and with very different kinds of brains. But the independent evolution of intelligence would be less surprising if consciousness was ubiquitous and ever-present. Then we could see how different lineages could adapt to it, just as they have adapted, with many different kinds of eye, to the universal presence of light.

Clearly, for day-to-day purposes, the notion of a universe led by consciousness is counter-intuitive. Yet it is perfectly logical, it explains a great deal, and it has occurred to a great many outstanding thinkers in every great culture over many centuries; and as a very considerable bonus, it has been finding significant support from thoroughly modern physics for the better part of a century. None of this means that it is necessarily true but it does mean – surely? – that it has to be taken seriously. This idea, which stands right at the heart of several thousand years of theological and philosophical tradition (and probably a great deal more before that that we don't know about) and is now at the heart of *avant garde* science, seems to kick ultra-materialism firmly into touch. By the same token it also lends support to the broad idea of transcendence – that there is more to the universe than meets the eye, and that this 'more' is far from trivial. If we are interested simply in day-to-day survival and in 'competing' in the 'global market-place', then ultra-materialism will suffice, at least in the short term. But if we are seriously interested in *truth*, as science ought to be, then the general idea of transcendence, and the specific idea of universal consciousness, cannot simply be dismissed. Alas! Because ultra-materialist attitudes seem to bring home the bacon in the short term, and because our world is focused more and more narrowly on immediate material advantage, ultra-materialism wins. But as we will discuss in more detail in the last chapter, the more we allow it to win, the more we dig our own graves – because whether we take a materialist view or a transcendent view profoundly affects our attitude to all of life, and hence to the practicalities of politics and economics. Demonstrably, crude materialism is very damaging indeed. If we persist with it, it surely will prove terminal.

Yet we can take heart – because more and more scientists are taking the phenomenon of consciousness very seriously, and the dialogue between scientists and theologians, which has never stopped and has often been very fruitful (despite what we have commonly been told) is now very rich. There's a growing literature by physicists and biologists, psychologists and anthropologists, theologians, philosophers, and other thinkers (as listed in the bibliography). Many feel (as I do) that we are witnessing a true 'paradigm shift' – the most significant since the Renaissance itself. To more and more people, ultra-materialism seems a busted flush. The hubris to which it gives rise – the idea that we can and should 'conquer' nature, with custom-built GM crops and all the rest – is presented to us from on high as ultra-modern, but in truth is seriously old-fashioned, rooted in the over-confidence of the eighteenth century.

Science doesn't have to be ultra-materialist. Those who insist that it is bound to be ultra-materialist simply don't know what science is, and what it isn't, even though they may themselves be professional scientists, and Fellows of learned academies to boot. Science education, alas, does not routinely include philosophy, and tends positively to shun theology, though firing many a barbed shaft, for the myth has it that religion over the centuries has simply got in the way of science, and held it back. The essential, unifying concept of metaphysics does not get a look in. But now the boot is on the other foot. The eighteenth century materialism that is now perceived to be modernity is halting humanity in its tracks, and indeed is threatening to kill us all.

The second area of science where scientists cannot avoid metaphysics – where metaphysics has crept in whether the scientists recognise the fact or not – is the one that is right at the heart of modern biology: evolution.

Does life have purpose? Does evolution have direction?

Those who think about evolution have grouped themselves in a variety of factions which like members of Christian denominations, glare at each other across no-man's land, and sometimes turn nasty. The most conspicuous disagreement – noisy, sometimes violent, but in general not very interesting – is between fundamentalist 'Creationists' who don't believe in evolution at all, at any price, and those who feel that evolution is a prime fact of life. For my part I am sure that evolution is the way of the world, and that it is a wonderful idea that has huge spiritual significance of a positive kind. But the vast army of Creationists spend their lives pointing out the real and imaginary flaws in Darwin's great idea – many if not most of which Darwin pointed out himself; while a small battalion of Darwinian biologists, some of them equally fundamentalist, spend their lives bashing the Creationists, and although the bashing has sometimes proved lucrative it has on the whole been most unedifying. I like the idea that was grasped by many a cleric in the immediate aftermath of *Origin of Species,* summarised not least by John Lennox, a professor of mathematics at Oxford and an active Christian: that it is perfectly logical to believe in a Creator *and* to believe in evolution. If you truly understand both concepts, then there is no conflict.

Among the people who accept evolution ('evolutionists') there is significant disagreement about the particular role and status of natural selection. Detractors point out that chance events – like the asteroid that finished off the terrestrial dinosaurs – may in practice play a bigger role than the gradual transformations that Darwin envisaged. Others point out that the course of evolution is restrained and directed by the machinations of the genes – including their constant Darwinian struggle to out-do each other. As in Greek tragedy, the real action takes place off-stage. Darwin himself said that he did not suppose that natural selection was the only mechanism at work – just that it is very important. The twentieth century English engineer-biologist John Maynard Smith pointed out that whatever else may be going on, natural selection is the only explanation that science can provide to explain how it is that living creatures are so *supremely* adapted to their individual circumstances. If nothing else, natural selection is the fine tuner; and in the end it's the fine tuning that counts.

The most important questions, however, are not whether evolution is a fact of life (as far as I am concerned, of course it is); or about the particular mechanisms behind evolutionary change. The questions that matter are of a metaphysical kind: matters in which science certainly has important things to say, but cannot hope alone to answer definitively. In this context three metaphysical questions are outstanding. Does evolution have direction? Is there any worthwhile sense in which evolution leads to progress? And, is there *purpose* in evolution?

The conventional, materialist answer to all three of these questions is 'No!' – sometimes accompanied by a snort of derision, for it is taken to be self-evident that only an idiot would raise such issues. Yet there is plenty of reason to think that in all cases the answer should at least be a cautious yes – at least up to a point. Certainly an outright 'No!' cannot be justified. Flat denial is nothing more than dogma – an opinion, a point of view, and the fact that it often emanates from scientists does not mean that it is good science. The discussions are endlessly complex as always but here, as I see things, are the main issues.

Is there progress in evolution?

Materialist biologists don't like the idea of progress in evolution – for three main reasons. First they think that the idea of progress implies value judgment – that the creatures that are deemed to have 'progressed' further are felt to be in some sense superior, and more worthy of respect. But the answer to that is a simple denial: 'No: progress doesn't imply value judgment'. I feel for example that in many technical respects flowering plants have progressed far beyond the level of mosses. But I have no sense that flowering plants are superior. (Mosses are one of life's delights).

Then the anti-progressionists suggest that the concept of progress makes no sense unless we suppose that each particular lineage is progressing towards some particular end. The idea that life has goals is called 'teleology' (from the Greek *telos* meaning goal) and was much favoured by Aristotle, but is an anathema to the materialist. But again, not so: the idea of progress need not imply any particular goal. It merely suggests, as is obviously the case, that the universe (or more specifically the Earth) poses a range of problems, which living creatures have to solve, if they are to continue living. Or we could say in more positive vein as modern business people like to do, that the universe (or specifically the Earth) offers a range of opportunities, which living creatures can exploit. All this is another way of saying what we said in Chapter 3 – that evolution can be seen as dialogue, between living creatures and their surroundings.

The fossil record clearly shows that as the generations pass, different lineages of creatures between them solve a greater and greater range of the problems, and exploit more and more of the opportunities. Thus in the beginning all creatures lived in water – and then lineage after lineage adapted to the land, and lineage after lineage after that adapted in many different ways an air-borne life – floating aloft or actively on the wing. In all cases, different lineages came to exploit each opportunity in more and more ways. Within each chosen métier lineage after lineage began to exploit the possibilities more and more adeptly. Beyond doubt, *Archaeopteryx* could fly – but everything we know about aerodynamics suggests that its flight was very limited. It was probably very stable in the air – in the manner of a paper aeroplane – but it could not manoeuvre. It must have used a great deal of energy as it flapped its way from tree to tree or from rock to rock – slowly but not too slowly or it would have stalled. Some modern birds are generalists just as *Archaeopteryx* was – yet many of them are also fast and/or manoeuvrable, but mechanically very efficient, as *Archaeopteryx* very obviously was not. Pigeons fit this bill. Other modern birds are supreme specialists – peregrines, hummingbirds, fulmars. All of them do things that *Archaeopteryx* could not, and with far less fuss. To suggest that there has been no progress over evolutionary time seems simply perverse. We might as well argue that modern aircraft, jumping jets and 747s and all the rest, have not progressed since the Wright brothers. Of course they have.

All in all it is perfectly reasonable to see the problems posed by the environment, and the opportunities it offers, as conceptual milestones: and lineage after lineage, as time goes by, pass those milestones. Or yet again: evolution emerges as a dialogue between life and the rest of the universe. Over time, the dialogue has grown richer and richer, with more and more threads to it, more and more finely nuanced. I would call that progress. It is rather like the progress from plainsong to symphony. The symphony may not be aesthetically superior, or more deeply felt, or, to many ears, more pleasing. But there is definitely more to it.

Finally, the anti-progressionists claim that the concept of progress is irrelevant. All that matters is survival – and non-progressive creatures may survive just as well, and often a great deal better, than the more progressive types. The answer to this is, 'Up to a point!' Simple mosses thrive in this world as well as super-sophisticated flowering plants; and since they evolved earlier, they have been around for far longer, and by that criterion we can say that they have proved to be *more* successful. But within each evolutionary niche, each modus vivendi, we repeatedly see more sophisticated creatures (as measured by the objective criteria of engineering) supplanting the more basic types – just as pigeons and peregrines and many more have replaced *Archaeopteryx*. For all kinds of reasons we cannot simply assert that the creatures that are measurably more able (swifter, cleverer, and so on) will survive, for as the preacher in Ecclesiastes reminds us, 'Time and chance happeneth to all'. Neither need we suppose, as discussed in Chapter 1, that the technically more able creatures did battle with the more primitive ones (where 'primitive' is a technical term meaning 'more similar to the ancestral form'). But the fact that the technically more able creatures are with us still while their primitive forebears are gone, suggests that there surely is some correlation between technical ability and survival. Indeed, if it were not so, it's hard to see why any lineage would ever evolve at all.

In short, taken all in all, I reckon that the idea that there is no progress in evolution is another example of *Zeitgeist* creeping into science. It is a manifestation of political correctness.

Does evolution have direction? Does life have purpose?

Does evolution lead each lineage towards particular goals? Many biologists who accepted Darwin's ideas were devout Christians, and remained devout -but this did raise a few problems. Notably, Genesis strongly suggests that God created each of his creatures ready-made, in their finished form; whereas evolutionary biology says that every creature first appeared in some primitive form and reached its present state only after many generations of change. One way to reconcile the two kinds of notion was and is simply to suggest that evolution is God's favoured method. He did indeed create the ancestors of each lineage and then perfected the lineages over time – but from the outset, he had the end-points of each lineage firmly in mind. This kind of thinking encouraged the idea of 'orthogenesis': the notion that evolution proceeds in straight lines, from primitive to modern, without deviation. In this vein, old-style textbooks showed horses evolving over fifty million years or so from the fox-sized, five-toed 'eohippus' (better known as *Hyracotherium*) to the snorting, prancing, one-toed jades of today. Similarly, human evolution has commonly been presented as a non-stop progression from a creature like a chimpanzee to upright us – as still featured in many a cartoon, not least on tee-shirts and car-stickers.

But as more and more fossils have come to light this past 150 years or so, it has become clear that all is not so simple. The evolutionary tree of horses, and of human beings, is not a straight line – a tree in the form of a cordon. It is a bush, with all kinds of deviations. There were several lineages of three-toed horses, evolving independently in parallel, and more than one one-toed lineage, and there was even an odd-ball in the Pliocene that went back to the woods and became small. Shockingly, many still feel, the human tree is bushy too. Through most of the past four million years up to half a dozen species of recognisably human-like species have co-existed. We, *Homo sapiens*, have had the planet to ourselves only for about twelve thousand years (which is when the last of the miniature island people from Indonesia, *Homo floresiensis*, are thought to have died out). The fossil record simply doesn't suggest a single-minded development from primitive prototype to modern perfection.

Add to this the idea mooted earlier – that lineages of creatures are not shaped detail by detail by natural selection, but are knocked this way and that or wiped out altogether by chance events such as asteroids – and it begins to seem as if evolutionary history can have no direction at all. Indeed Stephen Jay Gould argued in the 1980s that the course of evolution could be seen as a 'random walk'. Evolution at any one time could go any which way, depending only on the immediate selective pressures. What turned out at the end really was, and is, just a matter of chance. The impression of direction is delusory.

Yet, when we look more closely, the path of evolution seems far from random. Notably, as Cambridge palaeontologist Simon Conway-Morris in particular has pointed out, there are the phenomena of parallel and convergent evolution – thousands and thousands of examples of them. Different lineages of creatures that may be quite unrelated often live in the same kinds of habitats and exploit those habitats in the same kinds of ways and finish up looking remarkably similar. The small burrowing worm-eating mammals known generally as 'moles' offer a small but striking example. By 'mole', most of us mean the amiable little fellow of *Wind in the Willows,* the European mole; related to shrews in the group that is now called Soricomorpha (after the Latin name for shrew, *Sorex*). The soricomorphs in turn belong to the even grander group of the Laurasiatheria ('the beasts of the northern continent') – which also includes for example the bats, whales, cattle, horses, dogs, and cats. But there is a remarkably similar 'mole' in Madagascar that belongs to the tenrec family, Tenrecidae. The Tenrecidae is unique to Madagascar – and it belongs to a quite different mammalian group, the Afrotheria ('the beasts of Africa'); and the Afrotheria also includes the elephants, sea-cows, and aardvarks. Within the Afrotheria, too, is a quite different group of 'moles' known as golden moles – different enough from all other moles to be given their own family. For good measure, the marsupials, which include the kangaroos and the koala and are quite unrelated either to the Afrotherians or to the Laurasiatherians, have produced a very convincing looking 'mole' of their own. So although all moles look much of a muchness they in fact belong to four quite different families of which one is related to whales and bats and horses, two are related to elephants and manatees, and one is related to kangaroos. This is convergent evolution *par excellence* – yet there are many more examples including many that are even more striking.

Clearly, evolution cannot be seen *simply* as a 'random walk'. Nature clearly favours a fairly long shortlist of very particular designs – worm-like, fish-like, mole-like, bird-like, flower-like – and repeats them over and over again. If each creature is seen as half of the evolutionary dialogue – each a response to the problems the universe poses – then we can (in theory) reverse-engineer, and infer what kinds of problems the universe is, in fact, posing. Overall it looks as if the universe as a whole, of which the Earth is a manifestation, does *not* offer an infinity of possibilities, for creatures to exploit every which way. Clearly the universe favours some courses far more than others. To put the matter whimsically, it has its own ideas about how things ought to be done. Overall the universe seems to be working, gradually but steadily, through its own agenda. Indeed we can realistically see the evolution of the universe as an unfolding (which is what the word 'evolution' literally means). Over time, the universe unfolds to realise its own potential – to become itself. Here we might also invoke the concept of 'purpose'; and suggest that it is the purpose of the universe simply to become itself, over time; and evolution (including natural selection) is its method.

The way in which each creature evolves determines which of the many possible directions the universe in fact, takes: what bits of the possible agenda in fact become manifest. We are one of those evolving creatures and uniquely, at least among the creatures that we know about, we are able consciously to *choose* what course we want to take. Our choices affect the whole universe because they are part of the whole. We carry a huge responsibility. Yet we don't have commensurate power. We are not omnipotent, and never can be. In bearing our vast responsibility, humility is definitely in order.

Of course, ultra-materialists, scientists or not, will assert that this interpretation of evolution – the universe unfolding to become itself – is nonsense. But actually, it is a perfectly reasonable point of view – based on observation, and with true explanatory power. That too is the best that can be said for the opposite, negative, ultra-materialist interpretation – that evolution is just an opportunist romp, going nowhere. It is just a point of view, no more, no less. Ultra-materialist scientists tend to argue as if their particular points of view must be correct, because they emerge from science. But as we will see a little later, they argue this only because they are not good philosophers, and do not know what science really is, and what it is not. But we will come to that. First I want to address the last big metaphysical question that hangs over all scientific deliberation – the sting in the tail. Why should things be the way they are? In truth the question is not 'Why?' which has all kinds of connotations. Neither should we simply ask 'How?' The question that really matters (as Ludwig Wittgenstein put the matter) is 'How come?'

How come?

The question 'How come?' invites many kind of answer. For philosophers it implies, 'Why does the universe exist at all?' Theologians have often asked, 'Who made it?' – and again:

'Why?' Physicists have largely settled for the Big Bang – or that at least is the most favoured notion right now – but still they must ask: 'What were the preconditions that led up to the Big Bang?'

There are many mysteries. The concept of Gaia, discussed in Chapter 5, has shown more clearly than ever before that life is even stranger and more wonderful than it may seem. Strangest of all perhaps is that it should exist at all: so many minutiae must be in place if life is even to begin – and it is even harder to envisage a planet in which life can actually persist. Always it seems as if ecosystems must simply run down – and yet there is a host of mechanisms to ensure that they do not; and ecosystems in practice have gone on getting richer and richer for the past 3.8 billion years or so (at least they do if we include the concept of consciousness and choice of action among the components of ecology – as indeed we should). The universe as a whole, of which the Earth is the tiniest fragment, albeit a very privileged one, is equally unlikely. If any one of the many different particles that we know about, or any one of the fundamental forces, was ever so slightly bigger, or smaller, or didn't exist at all, the whole caboodle would fail. There would be no atoms, let alone anything else. The universe indeed is like Baby Bear's porridge: not too hot, or too cold, or too sweet or too salty, but 'Just right!'; and so the extraordinary 'rightness' of the universe has been called 'the Goldilocks effect'. So we have Gaia and Goldilocks – an extraordinary pair; just as hugely mysterious as the fact of consciousness itself.

A key problem, for physicists and theologians alike, is that of 'infinite regression'. So it is that many a theologian has sought to explain how it is that the universe began simply by suggesting that God made it. But this raises the question, 'Who made God?' Some ultra-materialist atheist scientists seem to think that they were the first to notice that this was a difficulty – and use it to deride the idea of the Creator-god. But ancient philosophers and theologians of all religions have always recognised the problem, and addressed it in many ingenious ways.

The truth is, though, that that this fundamental question is insoluble. Sometimes we just have to accept that we are stuck with mystery. Aristotle (perhaps in a spirit of resignation) simply suggested that everything must have a cause – so there must therefore have been a *first* cause, that was not itself caused: 'the unmoved mover'. Sixteen hundred years later Thomas Aquinas equated this 'unmoved mover' with God – and used the idea as part of his proof that God must therefore exist. Aristotle's idea was not that the unmoved mover could be shown to exist, empirically. He simply pointed out that such an idea is *logically* necessary. Aquinas was a great logician (as were all the great medieval theologians, contrary again to what we commonly told) and so he argued, in Aristotelian vein, that the existence of God is *logically* necessary. Of course the idea of an unmoved mover, or of God who created himself, is odd, and counterintuitive. But then, the fact that the universe exists at all, and must (surely?) have begun some time, and must (surely?) have had a cause, demands a counterintuitive explanation. The unmoved mover, or the self-created Creator, is, frankly, as near as anyone has got.

Some modern cosmologists will say, 'No it isn't! We have some perfectly good theories to explain how it all began!' Prominent among them these days is the multiverse idea. There could, the idea has it, have been an infinite number of different universes – and so perhaps there were. But in most of them, the forces and particles and physical constants didn't all fit together perfectly, Goldilocks style, and so most of these would-be universes failed. But at least one of them didn't fail – the one that now exists. Thus the multiverse theory attempts to explain the existence of all we see by a form of natural selection. There were many options on the primeval table, but only one of them worked; just as in the past there were many lineages of horse, or indeed of humans, and in the end only one of them came through.

The multiverse idea is ingenious but there are three obvious snags. First, there is no empirical evidence for it. It is merely thought to be in some way *necessary*, like Aristotle's unmoved mover. Secondly, it is hard to see how such an idea could be critically tested – but science qua science (as opposed to science qua armchair speculation) is supposed to deal only in ideas that can be tested and at least shown not to be untrue. Thirdly, it does not solve the problem of infinite regression – the problem we began with. For how did the multiverse come about? And why?

So why do scientists – or at least a great many of them, including some of the most eminent – embrace the multiverse while rejecting, say, the idea of the Creator god? If we wanted to be sceptical, we could say that both are ad hoc attempts to provide an explanation where none otherwise seems to be possible. Scientists of the hard-nosed kind will always insist that their ideas are 'robust', because they are founded on hundreds of years of observation and measurement and maths, while theologians merely dream. But the contrast here is not between a robust, cast-iron theory of science and an old-fashioned piece of story-telling. It's between two pieces of story-telling from different traditions. As we saw in the Introduction to this book, Richard Dawkins has famously written:

> The universe we observe has precisely the properties we should expect if there is, at bottom, no design, no purpose, no evil and no good, nothing but blind pitiless indifference.[32]

Richard Dawkins is an Oxford professor and a Fellow of the Royal Society and is widely perceived therefore to speak on behalf of science. But there is nothing 'scientific' about such a statement. Dawkins' perception of purposelessness is merely a point of view. It also seems to run contrary to all that we know about Gaia and Goldilocks.

So those who truly seek after truth are left with another very big question of a different kind. How should we distinguish between one point of view and another? How do we know what's true?

8. What's True? What's Good?

I was brought up with a myth. It wasn't written down in any formal, universally acknowledged text. We didn't chant it round the camp-fire. But it has been ever-present nonetheless, dinned into to us through all kinds of channels from textbooks to TV ads.

The myth presents us with a view of science as an edifice of truth, rising inexorably from the dust of ignorance. The stones of this edifice are *facts* – reliable and repeatable observations, measured this way and that, and the mortar that holds them together is scientific theory.

The edifice, thus constructed, is impregnable. Facts are solid: you can't go wrong with facts. As Thomas Gradgrind, schoolmaster, put the matter in Charles Dickens's *Hard Times:*

> Now, what I want is, Facts. Teach these boys and girls nothing but Facts. Facts alone are wanted in life. Plant nothing else, and root out everything else. You can only form the minds of reasoning animals upon Facts: nothing else will ever be of any service to them. This is the principle on which I bring up my own children, and this is the principle on which I bring up these children. Stick to Facts, sir! *(Hard Times,* Chapter 1.)

The theories of science, by their nature, cannot be quite so solid as the bedrock facts but they are up to the task nonetheless – framed with the upmost care and tested to breaking point under the greatest possible variety of conditions. All the artificers of this mighty scientific edifice keep an eye on each other to eliminate error. This is 'peer review'; the sternest scrutiny of all.

One day, the myth tells us, the edifice of science will encompass all there is. We are promised omniscience. We don't know everything yet, but if we keep up the research, then one day we will – and that day will surely come sooner rather than later. Even as I write these words, the Higgs boson is still on the front pages – and is dubbed the 'God particle'; the key to everything that is.

Out of science comes 'high technology'. We've had brilliant technologies of a kind ever since we were human – from stone axes to Roman aqueducts to galleons and stagecoaches – but until recently, all technologies evolved from craft. 'High' technologies are modern. 'High' technologies depend for their existence on many decades of formal scientific theorising and nowadays they include jumbo jets and genetic engineering and IT. As science makes us omniscient, so high tech will make us omnipotent. We can 'conquer' nature, alter and exploit any bit of it at will. We just need to fund the research. The more we spend the more we can do.

All this is perceived above all to be 'rational', for science is presented to us above all as the exemplar of rationality. Not to be rational, the myth has it, is to be 'irrational', which, as discussed in the introduction, is commonly taken to mean 'mad'. Clearly, then, anyone who questions the supremacy of science and the efficacy of high tech should be seen not simply as a back-slider, but as a potentially dangerous lunatic. People who object for example to GM ('genetically modified') crops and livestock are commonly cast in this light: misguided; subversive; standing in the way of human progress.

Yet, irony of ironies, the gung-ho view of science that is seen to be ultra-modern in truth is seriously old-fashioned. It belongs to the eighteenth century. I just don't see how anyone could persist in this mythology who has any cognizance at all of the philosophy of science that emerged in the twentieth century. In fact no-one who truly knows what science *is* could hold such views – even though many scientists do hold such views, or speak as if they do, including some scientists in high places.

For although science is truly a great creation, one of the greatest collective enterprises of humankind, it cannot properly be seen as an edifice of truth. Neither does it provide a royal road to truth. It is not the exemplar of rationality and in any case, rationality is not what it seems and it cannot lead us inexorably to truth any more than science can – not, that is, unless we define 'truth' very narrowly indeed.

So if we are seriously interested in truth, and are not seeking simply to tell ourselves a convenient story, then the myth needs to be examined and exposed. We need to ask what science really is, and what it is not. We need to explore the limits of rationality itself. In short we need to ask how we know what's true. But then, if we are really interested in truth, we need to ask how we can possibly hope to understand the universe and life and ourselves, if science and rationality can't take us to where we want to go. Which brings us back to metaphysics. So to begin at the beginning:

What science is and what it is not

In olden times 'science' just meant 'knowledge', which is what Aquinas meant by it when he declared in the thirteenth century that 'theology is the queen of sciences'. More and more, however, it came to be equated specifically with 'natural philosophy' – the study of the material universe and the natural world. We can trace the roots of natural philosophy deep into ancient times – alchemy, astrology, the arts of the apothecary, all manner of crafts, and indeed magic. Shakespeare's Prospero in *The Tempest* can properly be seen as a proto-scientist (and we may note in passing that Shakespeare and Galileo were born in the same year). Many ancient philosophers too, both Arab and European, thought seriously about the kinds of approaches that could lead us to robust knowledge, and they can properly be called philosophers of science; and some of them are still cited, not least the fourteenth century English friar William of Ockham, or Occam (of whom more later). Most scholars agree, though, at least in broad terms, that recognisably modern science did not properly begin until the seventeenth century with a succession of giants including Johannes Kepler, Galileo Galilei, Isaac Newton, Gottfried Wilhelm Leibniz, and René Descartes, ably abetted by possibly lesser but nonetheless outstanding thinkers such as Robert Boyle, Robert Hooke, and the biologist John Ray. This, the 'scientific revolution', marked the culmination of the Renaissance and carried us into the Age of Enlightenment – in which, in effect, we are still living. But the pioneers of the seventeenth century were still called natural philosophers. The term 'scientist' did not come on board until 1834 (coined by the English philosopher and polymath William Whewell).

Scientists come in all shapes and sizes and different scientists in different fields approach their work in many different ways; and scientists, of course, are not the only seekers after truth. So what do these multifarious scientists with their multifarious endeavours have in common, which prompts us to bundle them all together under the common umbrella of 'science'? And what makes science different from other formal disciplines, such as history or economics? In truth, the grand unifying idea is very simple (in principle) and has been around for a very long time: that the ideas of *bona fide* science are, above all, *testable*. The basis of all science that is properly called science, is 'the testable hypothesis'. So what is a hypothesis and what counts as a test?

The testable hypothesis

All science must begin with observation, some inkling of what is actually out there – with what Thomas Gradgrind called facts. Science is still rooted in natural philosophy after all, and its principal job is to explain how everything that is out there actually works. The hypothesis is the scientist's first formal attempt to provide the required explanation – and in essence, as the twentieth century particle physicist Richard Feynman put the matter, the hypothesis begins with nothing more nor less than a guess. It's a sensible guess, shaped by the scientist's own knowledge of what's what, and what's possible. But it's still a guess. We may feel that Feynman was simply being mischievous, for he was known for his irreverence – but in truth, nearly three hundred years before Feynman, Isaac Newton said much the same:

> No great discovery was ever made without a bold guess.

The preliminary guess is then refined into a formal hypothesis – the first draft of an explanation; and then this formally framed hypothesis is well and truly put through the mill. The standard way to test a hypothesis is to predict what would be the case if the hypothesis was actually true – and then make further observations, or preferably to carry out experiments, to see if those predictions do actually stand up. Throughout the whole process – gathering the initial facts, framing the hypothesis, and analysing the results of experiments – maths in various guises plays crucial roles. Indeed the outstanding scholar of Chinese science, Joseph Needham, was wont to define science as 'the ruthless mathematicisation of ideas'. Only when a hypothesis has been beaten this way and that by a lot of people over a long time, and stood up to the tests, is it admitted as a 'theory'. 'Theory' in science does not mean what it means in common parlance – a whimsical top-of-the-head notion. It's an idea that has truly been tried and tested (and the tests almost unconditionally must in part be mathematical).

All this adds up to what is called the 'method' of science – and it is, beyond doubt, extremely powerful. To some extent the myth is true. The gathering of facts these days is assiduous in the extreme. In general, no observation is admitted as a 'fact' to be taken seriously unless it can be made repeatably and reliably, more or less to order, preferably by many different, independent observers, and quantified. Any hypothesis that is considered to be in any way important is tested and re-tested in a hundred different ways in many situations – and the results may then be analysed by several different statistical protocols. Peer review is fierce. Scientists can be very hard on each other. The occasional cheats, once exposed, are hounded without mercy.

By the eighteenth century, within decades of the birth of recognisably modern science, the myth of potential omniscience was already established. For although the proto-scientists like Prospero had dreamed of universal principles that truly would explain how life and the universe work – 'nature's secrets' – they also had reason to fear that there were no universal principles of a reliable kind: that the universe in the end is whimsical and forever beyond our ken. But by the eighteenth century it seemed that such fears were unfounded. The seventeenth century scientists and Newton in particular established the idea of the natural 'law'. Newton indeed envisaged the universe as a giant exercise in clockwork – and nothing is more orderly or reliable than a well-made clock. Thus inspired, the Enlightenment mathematician and physicist Pierre Simon de Laplace (1749-1827) declared in the early nineteenth century that if only we knew where everything in the universe was at any one time, then we could predict with certainty how everything would turn out for the rest of time. It's reported that when Napoleon asked Laplace where God fitted in to his grand scheme of things he replied, '*Je n'avais pas besoin de cette hypothèse-là.*' ('I had no need of that hypothesis.'). This tale is apparently apocryphal – and in any case Laplace did not, as it may seem, intend to deny the existence of God. He merely meant to indicate that once God had installed the laws of physics, He had no further need to intervene (which is a version of theology known as 'deism'). Confidence in the laws of physics, and in our ability to understand them, continued to grow; so much so that the German American physicist Albert Abraham Michelson declared in the late nineteenth century that theoretical physics was more or less sewn up. It just needed a few more decimal points to refine the equations, and then we would understand more or less everything about the material universe that there is to understand.

But it didn't quite work out that way. At least, science overall in the twentieth and early twenty-first centuries has a curious ambivalence. On the one hand confidence has increased – both in the power of science to explain, and in the power of high tech to control. It's a prime conceit of our age, a significant component of the *Zeitgeist,* that human beings have decisively replaced God, both in theory and in practice. Yet is seems to me that the history of twentieth century science, and of the philosophy of science, must undermine such confidence absolutely. Although Michelson was an outstanding scientist (together with Edward Morley he measured the speed of light and was the first American to receive the Nobel Prize) his faith in late nineteenth century physics now seems ridiculous. For his own work, and Morley's, helped Einstein to formulate his ideas of relativity, which turned all physics on its head. For Einstein showed that Newton's mechanical laws are seriously limited. They are very useful – they provide the basic physics needed to send rockets to the moon – but they describe only one small aspect of reality. Roughly at the same time as Einstein described relativity, Max Planck's early experiments on the radiation of heat led remarkably quickly to quantum physics – and that revealed a world that is seriously weird in which Newton's laws do not seem to apply at all. So within a few decades of Michelson's hubristic declaration, all physics was up for grabs.

Despite all this the confidence seemed to increase – boosted not least by the philosophy known as logical positivism, which was developed by a small group of intellectuals in Vienna in the 1920s.

Logical positivism: a twentieth century diversion

Logical positivism had an aim which seems commendable enough, and has always been a *leitmotiv* of philosophy: to find a *basis* for truth that is absolutely reliable. It was this that led Descartes to his 'I think therefore I am': the fact that he thought, he said, was the only thing he could be certain of. For the logical positivists, the thing to cling to was the principle of 'verifiability'. They put their faith in whatever could be pinned down – and in nothing else. They considered no statement even to have *meaning*, unless it could be verified. By the same token, it was ridiculous even to ask questions of a kind that could not be answered by verifiable statements. So for example the question 'Does God exist?' cannot even in principle be answered definitively – and so, said the logical positivists, it was nonsensical. The concept of God was not simply elusive; it was literal gibberish. No idea that is not verifiable could be admitted at all.

So what ideas were 'verifiable', in the opinion of the logical positivists? Only those based on repeatable observation, and measurement, and mathematical analysis. In short, only science was verifiable; and therefore, only the assertions and ideas of science even had meaning. Indeed, a comment by one the founders of logical positivism, Rudolf Carnap, anticipated Peter Atkins by more than half a century:

> ... there is no question whose answer is in principle unobtainable by science.[33]

Everything else – again including all of metaphysics and most of philosophy and of course all religion – was junk; not even worth thinking about.

Logical positivism was hugely influential throughout the mid-twentieth century, promoted in particular by the very forceful and eloquent Oxford philosopher A.J. Ayer. Along the way it helped in particular to justify the science of behaviourism as outlined in Chapter 4, and it gave a significant boost to hard-nosed materialism and atheism in general. It finally began to peter out after about 1970 – but, very clearly, at least in subliminal form, it still lingers on in the heads of a great many materialist-atheists. For many scientists in high places and for their admirers and hangers-on, science in the guise of hard-nosed materialism is the only game in town. In fact we can trace a clear line of thought from the eighteenth century rationalists through Carnap to Peter Atkins and, I would say, to Richard Dawkins and Dan Dennett (with Thomas Gradgrind waving from the sidelines).

Yet throughout the twentieth century, the foundations of hard-nosed materialism were undermined.

A change of direction

First, as we have seen from earlier chapters, new discoveries in science itself soon began to show that the confidence of the Enlightenment (as exemplified by Laplace) was seriously misplaced. Quantum physics shows us that the behaviour of the most fundamental particles of which the universe is composed is anything but predictable. Electrons and photons and all the rest leap from place to place apparently entirely at random. We can and do make broad-brush predictions despite this, but detailed crystal-gazing is absolutely off the agenda. On a different tack, although not entirely unrelated, is the principle of non-linearity, as cited throughout the earlier chapters and especially in Chapter 4. In nature, cause and effect depend on so many different factors, of which many are bound to be unknown (though we have no way of knowing how many are unknown). So there can be no certainty because the universe itself is uncertain; uncertainty is built into its fabric.

There is also a brief but devastating catalogue of theoretical reasons why science *cannot* be the edifice of irrefutable truth, or provide any kind of royal road to truth. First, the Viennese-British philosopher Sir Karl Popper re-examined and refined the idea of the testable hypothesis. He declared, first, that no idea about the empirical universe can, in reality, be shown beyond all reasonable doubt to be true – so the logical positivist demand that all bona fide ideas must be 'verifiable' seems to go out of the window. For example, he said, you cannot prove beyond doubt the hypothesis that all swans are white because no matter how many you find, you can never be sure that there isn't a black one somewhere; so the hypothesis, 'All swans are white', cannot be verified. But if you found one black swan (and there are plenty in Australia) then you would instantly *disprove* that hypothesis. In short, said Popper, we should not simply demand that an idea of bona fide science should be testable. Neither should we demand that it must be verifiable – for that is not possible, beyond all possible doubt. But it should be 'falsifiable', at least in principle. If we take

Popper's idea to its logical conclusion (as some have been inclined to do) then 'scientific truth' emerges as the sum total of all the testable hypotheses that have not yet been shown to be false.

To be sure, many philosophers have questioned the principle of falsifiability from various angles but it seems, nonetheless, to deliver a very severe blow to logical positivism. For how can we refuse to admit any idea that cannot be verified beyond doubt if *no* idea can ever be verified beyond all doubt? And where does this leave the edifice of irrefutable scientific truth? In general it seems that all the ideas of science must be considered provisional, waiting to be knocked off their perches by some new observation that cannot quite be fitted in to the existing theory. Often as science has progressed (and the joy of science is that it really does progress) we have seen old ideas thrown out, sometimes after decades or even centuries of acceptance, as new observations come along. Often, however, old theories are simply subsumed within larger ones, just as Newton's ideas of motion and mechanics were subsumed within the ideas of Einstein. But even Einstein's great vision of relativity must be considered provisional. (After all, Einstein's relativity can't yet be squared with the ideas of quantum physics, so a larger theory is clearly needed to embrace them both).

Yet Enlightenment-logical positivist confidence was dented even further, or it certainly should have been, by the insight of Kurt Goedel – who seriously questioned the supremacy of maths. Maths in science is taken to be the ultimate arbiter of truth. Scientific hypotheses are not considered truly to be respectable unless they can be subjected at least at some point to mathematical analysis. Darwin regretted his own lack of maths (and probably had too much faith in the insights of his mathematical cousin, Francis Galton). Pythagoras suggested in effect that God was a mathematician: that He devised His universe along mathematical lines. Plato followed Pythagoras in this, and he after all is commonly seen to be the founder of western philosophy.

But Goedel showed (and there is no doubt about this: all mathematicians accept his argument) that maths in the end, like all human endeavours, has a quality of arbitrariness. Some mathematical statements are simple tautologies, as in 2 + 2 = 4, where four is defined as twice two, and two is defined as half of four. In this there can be no worthwhile argument. But, said Goedel, any mathematical statement that isn't simply a tautology is bound to contain at least one assumption that cannot itself be proven. So maths is not the objective arbiter, with a status that seems to be semi-divine. It is revealed as a human invention, subject to human frailty like everything else that humans think about (and as the sceptics say, 'It fits reality where it touches').

Then there is the deceptively simple comment by the mid twentieth century zoologist Sir Peter Medawar – that 'Science is the art of the soluble' (see essays in *Pluto's Republic,* 1984). Medawar derived his title from Bismarck's comment that 'politics is the art of the possible' – and very apt it is. So it is for example that people throughout the twentieth century have been wont to ask why experimental psychologists have wasted so much time watching rats in mazes when what we really want to know is how the human mind works, and why we are beset by so many weird ideas and moods. One very good reason, said Medawar, is that it is *possible* to watch rats in mazes in a very critical way and measure their responses and draw statistically valid conclusions; but it is very difficult indeed to think up experiments that critically test hypotheses that relate to the human mind. As we saw in Chapter 4, consciousness, human and otherwise, remains a vast unknown.

Finally there's the simple and obvious caveat that the philosopher J.S. Mill drew attention to in the nineteenth century. We can never know everything there is to know; and therefore we can never know all the factors that ought to be taken into account in any one situation that we choose to study. At least, if we did know everything that was relevant, we could not know that this was the case. We could not tell how much we didn't know unless we were already omniscient, and so could measure present knowledge against all possible knowledge. As Donald Rumsfeld famously observed, though he was not noted as a philosopher, there are 'unknown unknowns'.

Even that, though, is not the end of the matter. For part of the myth of materialist science is that science is above all 'objective'. Scientists have often seen themselves, or been encouraged to see themselves, as 'pure' thinkers. In reality, though, scientists remain human beings and their researches are subject to the same human frailties that beset all of us. Darwin, as outlined in Chapter 2, was influenced by the unfolding history of his own times, by the general thinking of the Enlightenment and the particular input of Malthus, and the result of these influences is evident in *Origin of Species.* The behaviourist agenda was in large part a reaction to what early twentieth century psychologists saw as nineteenth century romanticism. Science in general and behaviourism in particular formed a positive feedback loop with logical positivism.

Similarly, in our own time, the controllers of world agriculture are besotted by biotechnology and in particular by genetic engineering, which they see as the answer to all humanity's food problems. In expensive brochures and official reports they seek to demonstrate that their love of GM is driven by an 'objective' assessment of all the relevant facts, which leads us inexorably towards high tech. In fact it is no such thing. Demonstrably, it is driven by the economic dogma which says that all farmers everywhere must strive to be as profitable as possible, in competition with everybody else; and that industrialisation (which above all reduces labour) is the only viable road to profit (so long as oil is cheap); and GM in various ways lends itself to industrialisation and to centralisation of power and hence of visible profit. The decision of various governments including Britain's to put their weight behind GM is presented to us as good science, and some of the people involved, both scientists and non-scientists, clearly believe that this is so. But in truth, the GM agenda is driven by economic dogma.

None of this is intended to belittle science qua science. The rigour of its methods is truly impressive. So is the quality of thinking that frames and tests the hypotheses. So are many of the scientists themselves – many combining enormous intelligence with personal humility and humanity. More even than this: science has shown us (at least as far as it is able to do so!) how wonderful the universe really is (which in some scientists at least, has reinforced their sense of transcendence). Finally, as a not inconsiderable bonus, science is useful. We, humanity, are probably at a point in history where we would find it hard to live tolerably without the high technologies that science so obligingly provides. Almost certainly, we could not live in such numbers.

But still science is limited. It does not tell us all there is to know, or what we might reasonably want to know. All its ideas in the end are provisional – because it is in the nature of science that its ideas can theoretically be disproved, or at least shown to be inadequate, and so are always waiting to be improved upon. Its ideas must always be partial, because science can deal only with bits of the whole – the bits that it is convenient to deal with. Scientific theories can never provide the whole truth – or even if, by some miracle they seem to do so, we could not know that it was the whole truth because we cannot know how much we don't know. All in all we might sceptically suggest that scientists seem to give such precise and convincing arguments to life's problems only because they take such care to tailor the questions, and leave whatever looks too hard off the agenda.

All in all, then, the model of science as an edifice of truth, an impregnable fortress, is nonsense. In so far as it is an edifice, said Popper, it is like Venice: impressive to be sure, yet founded not on bedrock but on stakes driven in to the mud. Or to shift the metaphor yet again it might be compared to a landscape painting, worked by a thousand hands, a never ending work-in-progress. Every now and again some extra brush-stroke, some extra observation, throws the whole picture out of kilter and then the scientist/artists have to begin again. Then we have what the American philosopher Thomas Kuhn in the 1960s called a 'paradigm shift' – a change of worldview. Whatever way we put it, all human understanding in the end is a story, a narrative; and that is as true of science as of everything else. Science tells a good story, beyond doubt, but to suppose that its story is the truth, the whole truth, and nothing but the truth is a grievous error; and as we can see from the present state of the world, it is a very dangerous one.

But where does that leave us in our search for truth – if truth, indeed, is what we seek? As I see it, we have three options. We can simply give up, acknowledge that the universe is beyond our ken, and get on with our lives. This is what most people seem to do and are obliged to do and at one level it is fair enough – but in the long term it is dangerous. It means that we leave the ideas that run our lives – especially those of politics and economics – unexamined. When things go wrong, as they very spectacularly do, we do not challenge the deep ideas in which our strategies and policies are rooted; or not, at least, with the rigour that they need to be challenged.

Or we can do as the logical positivists recommended, and define truth exclusively in terms of what science can tell us. Then we can simply dismiss all ideas that do not emerge from science. But that, as we have seen, is to accept as truth an account that we know, for the most fundamental reasons, is at best partial, and in various ways is flawed.

But there is a third option. We can accept that science is wonderful, and that rationality is vital – but then acknowledge too that if we are truly to get to the truth of things (or at least to get as close as possible) then we must venture beyond science, and beyond what is conventionally called rationality. We must engage the range of faculties that I suggest can reasonably be called intuition.

The absolute importance of intuition

Intuition, like consciousness (or intelligence, or mind, and all such grand ideas) is an elusive concept. Here, I am taking it to embrace the whole gamut of instincts and feelings which tell us, 'in our bones', what is actually the case; what we feel 'deep down' is right. When buses approach at speed we jump out of the way – and that is a instinct which I would say is a primitive form of intuition. At the other end of a long spectrum is the feeling for transcendence, including the idea of universal consciousness with the possibility of purpose and direction. The materialist-rationalist-atheists' dogmatic insistence what we can see and measure is all there is, is also in the end a matter of intuition, though the materialist-rationalist-atheists are loath to recognise the fact. In the end, more generally, the things that we take to be true are not what mathematical equations tell us is true, or even what seems to be the evidence of our own eyes. In the end, what we take to be true is what our intuition, our bone feeling, tells us is the case.

Is this mere whimsy? Not according to some of the world's greatest scientists (including those who claim to be among the most hard-nosed). Where, first of all, do the guesses come from that lead to formal hypotheses? Thinking plays a part of course but all great scientists (like all creative artists) rely very heavily on their unconscious minds to do the work. The nineteenth century German organic chemist Friedrich August Kekule envisaged the benzene ring (six carbon atoms in a circle) when he dreamed of a snake that held on to its own tail. Geneticist Barbara McClintock used to speak of the need to acquire 'a feeling for the organism'. Her method was above all to observe, then go for long walks and sit under trees and wait for inspiration. She marveled that although it was possible – indeed usual – to see the answers to the knottiest problems in a flash, it could then take a couple of hours to explain the idea even to well-informed colleagues. As she said in her acceptance speech in 1983 when she was awarded a Nobel Prize for her work on 'jumping genes':

> When you suddenly see the problem, something happens that you have the answer – before you are able to put it into words. It is all done subconsciously.

Newton too – according at least to legend – was inspired by trees. He is said after all to have arrived at his grand idea of universal gravity when an apple fell on his head. Whether it did or not, this putative apple did not come with a label, 'I have been subjected to a universal, mutual attraction between myself and the Earth – a manifestation, if you will, of the mutual attraction that exists between all objects that have mass'. That grand unifying idea came to Newton just as jumping genes came to Barbara McClintock, as a flash of inspiration, and no-one (certainly not Newton) knows how.

When such a grand idea has dawned, how do scientists – or any of us – judge that it's true (and that other ideas are false)? Again, it's a bone feeling – one that is clearly linked to our sense of aesthetics. 'Beauty is truth, truth beauty', said John Keats, which expresses the idea admirably. Or as the great English mathematician and physicist Paul Dirac put the matter, 'It is more important to have beauty in one's equations than to have them fit experiment'. Even James Watson, a notably hard-nosed atheist and defender of materialist science, has expressed this kind of sentiment. He *knew* the double helix model of DNA must be right, he said, because 'Something so beautiful cannot be wrong'.

The materialist-rationalist-atheists are wont to say that any idea that is arrived at by non-rational means is 'irrational'. But this a misuse of language: the antonym of 'rational' in this context is not 'irrational', but 'non-rational'. 'Irrational' implies 'mad' – unable to think straight – and therefore bad. 'Non-rational' simply implies thinking in ways that may deviate somewhat from the rules that are deemed to define rationality. But it was not strictly 'rational' to infer the benzene ring from the dream-world snake, or the jumping gene from the multi-coloured corncob. Yet it certainly wasn't 'irrational'. It was simply non-rational – a matter of intuition, including a sense of truth as beauty. Science itself could not proceed without such intuition. Without intuition we are forever stuck in the barren conceptual landscape of Thomas Gradgrind.

So instead of rejecting the non-rationality of intuition, we should embrace it. We should seek, actively, to cultivate it. Since intuition is demonstrably necessary in all walks of life – certainly when we ponder matters of morality, as discussed in the next chapter, and also when we ponder the nature of the material world – it is *rational* to cultivate intuition. If we assume that it is rational to seek truth (since truth is surely preferable or at least should be safer than untruth), and if we acknowledge that intuition may operate in ways that are non-rational, then it follows by Aristotelian logic that it can be rational to think non-rationally. Contrariwise, it is irrational – meaning mad – to dismiss our intuitions, our bone feelings, as decisively as the materialist-rationalist-atheists are so often inclined to do.

In truth, the tension between those who prefer to stick as rigidly as they can to the acknowledged rules of rationality, and those who are inclined to follow their gut feelings, the intimations of their bones, seems to be as ancient as human thought. It runs right through the culture of the Ancient Greeks – the 'Apollonic' rationalists on the one hand, and the 'Dionysiacs' on the other, following their feelings. We see it at the end of the eighteenth century and the early nineteenth, when the 'Romantic' poets, painters, and then musicians, reacted against the rigid rationality of the Enlightenment. What Samuel Taylor Coleridge and others called 'imagination' can certainly be seen as a cultivated form of intuition.

Most profoundly, we see the dichotomy of approach among philosophers and theologians – indeed among metaphysicians: the people who seek most directly both to find out what is true, and to ask how we know what is true. The difference was epitomised by a conversation in Andalusia at the end of the twelfth century between the Muslim teachers Ibn Rushd, known in the west as Averroes, and the young Ibn 'Arabi, as described by Peter Coates in *Ibn 'Arabi and Modern Thought*. Basically, Averroes was the arch-exponent of 'speculative thought'; while Ibn 'Arabi seeks what Averroes calls 'divine inspiration'. Averroes, already established as a great philosopher, perceives that Ibn 'Arabi's approach (which I suggest was a supremely refined form of intuition) is indispensable. Ibn 'Arabi is generally seen to be a Sufi, in the tradition of Muslim mysticism, and as I will argue in Chapter 9, the refinement of intuition is one of the keys to what it means to be religious.

What is the nature of intuition? All psychologists now acknowledge (again, in sharp contradiction to Descartes) that most of what we call 'thinking' is not carried out consciously. The primary role of our consciousness, it seems, is to access and to some extent to monitor the unconscious or subconscious thinking that goes on behind the scenes. Clearly, this subconscious thinking does not simply follow rules of logic, or indeed the rules of narrative of the kind that we expect to find in literature. The subconscious hops from notion to notion as manifest in dreams, or indeed in what William James called 'the stream of consciousness' (though it should surely be called the stream of unconsciousness). Then the flashes of intuition that led Newton to gravity and Barbara McClintock to jumping genes can simply be seen as a happy synchronicity between the conscious search for an idea and the subconscious quasi-random casting around.

But if we perceive consciousness to be a universal property of the universe (as discussed here in Chapter 8) and if we say that thinking creatures do not create consciousness out of nothing, but instead partake of what is already out there; then we can take a different view of intuition. Then we can see it as the mind's attempt to gear itself to the universal consciousness; in effect, to read the mind of the universe. You may feel that this is altogether too fanciful, a step too far. Yet the idea of universal consciousness is perfectly reasonable – not particularly counter-intuitive, and arrived at by various independent routes including that of modern physics; and so is the notion that thinking creatures partake of universal consciousness. So the idea that the unconscious mind tries purposefully to tap in to universal consciousness may be seen as a reasonable extrapolation, and not a particularly bold one. As discussed in the next chapter, I am sure that this is what mystics (including Ibn 'Arabi) aspire to do: to by-pass normal ratiocinative processes and tap in directly to universal consciousness. If we accept Peter Russell's suggestion that universal consciousness in effect means 'God', then to fuse the mind with universal consciousness is indeed to meet God.

But the out-and-out materialist-rationalists have one more shot in their locker. First, strange though it may seem, they are wont to appeal to the wisdom of the fourteenth century Franciscan friar, William of Occam. For Occam famously commented: *'Non sunt multiplicanda entia praetor necessitatem'*, meaning 'Entities are not to be multiplied beyond necessity'. Occam proposed this in response to the cogitations of Thomas Aquinas, who in the thirteenth century had sought to reconcile Christian theology with the ideas of Aristotle; and Aristotle had invoked the idea of 'essences' – that dogs or tables or trees each have an 'essence': a common thread of dog-dom or table-dom or tree-dom that unites each thing of any particular kind. Occam said 'We already have the objects themselves – dogs, tables, trees, whatever – so why do we need to invoke the additional idea of essences? Essences are clearly superfluous. So – *Non sunt multiplicanda entia praetor necessitatem!'*

Occam's adage has become a guiding principle in modern science – wherein it is known as 'Occam's razor' (although Occam himself never used that expression). But its scope has been extended to apply to all contexts, and its meaning has been modified. Most people nowadays seem to take it to mean 'Don't make ideas any more complicated than they need to be'. This seems perfectly reasonable – although we should remember Einstein's *caveat:* 'Everything should be made as simple as possible, but not simpler'.

But the idea has been expanded even further and a great many people nowadays seem to think that Occam's razor means: 'The simplest idea is the most likely to be true'. With this at least at the back of their minds, materialist-atheists are wont to argue that their own unyielding view of the world (there's nothing in it except atoms and the void) is more likely to be true than any idea which includes transcendence, because, at least on the face of things, the straightforward materialist idea is simpler. So it is that materialist-atheists tend to insist that anyone who argues for the existence of God should provide direct and special evidence for His existence. The idea that the materialist-atheist should provide cast-iron evidence for His non-existence is deemed to be out of order. The thinking seems to be that to claim that something exists is to make a positive claim, and that positive claims need special substantiation. To claim that something doesn't exist is merely to make a negative claim – and negative claims, for some reason, don't need such backing up. They are simply taken to be the default position. In line with the (seriously modified) version of Occam's razor, it is taken to be obvious that a statement that is a simple denial is more likely to be true than any positive assertion – so the positive assertion needs special proof. Occam himself, of course, as a devout cleric, did not believe that the idea of God was unnecessary. He took it to be self-evident (as many later philosophers did, including Descartes) that nothing could possibly exist *without* God. But he is invoked to support atheism nonetheless.

There is a lot wrong with this mis-use of Occam. First, to my knowledge, Occam was not suggesting that his principle provided any kind of royal road to truth. He merely proposed it as a means to keep our ideas tidy. He would surely have agreed with Einstein, that babies should not be thrown out with the bathwater. To suggest therefore that atheism is preferable to any sense of transcendence simply because it is simpler is to invite the response, 'So what?' The point is to find out what is *true*; and we cannot just assume that the ideas that seem simpler are, in reality, more true.

But then again: is it really legitimate to assume as the atheist-materialists do that the atheist-materialist position really is simpler? Does it really represent the default position – the idea that should be assumed to be true until proved not to be? I think not – for the idea that there is no God; or, more generally, that there is no intelligence or consciousness behind the universe – is itself a very positive assertion. It puts a very heavy burden on the material particles and the fundamental forces of which, the materialists suggest, the universe is entirely composed. It says that these particles and forces are such that *even without any assumption of driving intelligence*, they can gather together, and interact, in ways that produce galaxies and chemical elements, and living creatures, some of which are conscious, or at least have the illusion of consciousness, which at least suggests that they are capable of illusion. How can the atheist-materialists be so sure that the insights of materialist science, brilliant as they are, provide all the explanation that is needed? Or that science qua science will tell us all there is to know, probably soon, if only we keep plugging away along the approved lines of investigation -- observation and experiment and maths?

Yet the materialist-atheists are confident – or so they seem. Peter Atkins, Richard Dawkins, James Watson, and Steven Weinberg are among the luminaries who have insisted, often stridently and with contempt for all other points of view, in print and lectures, that theirs is the only kind of explanation that any truly sane person could entertain. Here, though, lies a deep irony. For although they claim that their confidence is rooted in reason, reason suggests that they have no right at all to be so confident. There are very good reasons to suppose that there *is* an intelligence behind the universe. To be sure, theologians and others continue to ask where this intelligence – this consciousness – lies. Is it part of the fabric of the universe; or separate from the universe (as suggested by the traditional Christian concept of God); or both (as suggested not least by Spinoza)? But there is, and always has been, broad agreement that this driving intelligence, this universal consciousness, must exist; and this (or something very like it) is what is meant by transcendence (or at least, that is what I am suggesting transcendence could mean).

But the materialist-atheists continue to claim that their view of the universe alone is 'rational', and must therefore be right; yet their adherence to materialism and atheism is again, in the end, a matter of intuition. They feel in their bones that atoms and the void are all there is, just as many (most) people feel in their bones that there must be more to it. In effect, then, the materialist-atheists claim that their intuition is superior to that of most other human beings, But I can't see why we should trust the intuitions of some other person, even if that person is a professor or a Nobel

Prize winner, if their intuitions run counter to our own. At least I can see why we should allow ourselves to be persuaded to abandon our own intuitions if the person with different intuitions provides superior arguments. But as yet, I haven't heard such arguments. I have only heard assertions, supported by appeals to techniques that are clearly inadequate, and by appeals to authority. This is ironical, given that the hard-nosed rationalists are apt to condemn all religions for relying do heavily on the authority of prophets.

Of course, our intuitions may sometimes lead us astray. For instance, the same primitive intuition that tells us to flee from whatever is dangerous and to leap from the path of runaway trucks has also left a great many people with a fear of snakes or spiders that far exceeds their dangerousness, On the whole, though, our intuitions serve us very well indeed. They are, after all, evolved, and if we see evolution as a dialogue, then we may reasonably infer that our evolved intuitions for the most part reflect the way the universe actually *is.* Certainly, there is no good reason to suppose that our inherited beliefs are likely to be wrong. There is no good reason (and I stress *reason*) to reject our intuitive sense of transcendence – the feeling that behind the appearances of things, there is an intelligence at work.

Still, though, some will feel that we should not lean too heavily on our intuitions to tell us what the universe is really like, and what is true. But for the third key question of metaphysics – 'What is good?' intuition is paramount.

What is good?

I was on a committee recently when a chap who was otherwise sensible said that we shouldn't discuss morality because it is simply a matter of personal opinion. That's a common view – particularly, perhaps, in our present age when everything, even morality, is supposed to be determined by the market. Apart from the odd taboos (which mercifully include human trafficking and child pornography) everything that people are prepared to pay for is deemed, at least in some circles, to be acceptable. I have even heard human cloning defended on the grounds that there would be a market for it.

The key point is as David Hume said – that we cannot arrive at a truly convincing moral position simply by thinking about it. In the end, morality must be informed by our feelings – by what Hume called 'passions', and I am happy to call 'intuition'; in particular, the bit of intuition that we call 'conscience'. Conscience may be hard to pin down but we all know it when we are struck by it – the feeling of uneasiness when we know we have done something wrong. This feeling seems far from arbitrary. At least, most people most of the time feel uneasy about the same kind of things – cheating; lying; spite; betrayal. The only people who don't feel uneasy when they succumb to such unpleasantness are properly called psychopaths. Either that, or they have in some measure been brutalised.

In practice, moral philosophers have formally classified codes of morality under three headings. The one that seems the most 'rational', and which was the most recently framed and now prevails, is commonly called 'utilitarian'. Moral good is judged by results; and the result that is desired is human happiness. So in utilitarian ethics a 'good' act is one that brings 'the greatest happiness to the greatest number'. Economists these days are wont to argue that what makes us happy is more material goods, and since economists rule in our market economy, what is deemed 'good' is what produces the most material gain for the least effort. In other words, 'good' tends to be defined these days in terms of cost-effectiveness. In modern agriculture, the most hideous cruelties and injustices are defended on the utilitarian grounds that they are profitable.

Most damningly, utilitarianism *per se* does not ask who is made happy by what – yet this is surely pertinent. A whole ring of paedophiles may be made happy by the abuse of one child – great happiness for a large number while only one is made miserable – but does that make it good? Of course not. We feel in our bones that it is vile and our bones are surely right. But utilitarian thinking alone cannot tell us *why* paedophilia is vile. Utilitarianism, in short, has its uses, but as an overall guide to life it is far too crude.

The second approach is called 'deontological' from the Greek *deon*, meaning duty. Doing good is doing what some acknowledged authority – God; the law; the boss – requires us to do. But the law can be unjust and bosses can be evil: war criminals who claim that they were 'only obeying orders' tend to be given short shrift. It surely is right to obey God – but not everyone believes in God. Those who do forever face the question – how do we know what God wants us to do? In practice people at large have relied on prophets – but the Bible itself warns us that prophets may be 'false'; and the world is full of cults, whose leaders often seem deeply suspect, at least to outsiders. So mere deontology doesn't seem to work either.

The third approach is that of 'virtue ethics' – first outlined in the western tradition by Aristotle. Virtue ethics focuses on *attitude* – to ourselves, to other people, to other creatures, and to the world at large. Attitude, after all, determines everything else that we do: whether we treat other people as potential enemies or potential friends; whether we treat a forest as a sacred place, or as a resource.

Religious moral codes are commonly thought to be deontological. After all, the three Abrahamic religions (Judaism, Christianity, and Islam) all lean heavily upon the Ten Commandments – and if commandments are not orders from above, then what are they? Yet the commandments seek largely to define attitude: worship God; honour your parents. The great prophets and religious sages stress attitude above all else. Jesus spoke of love, Muhammad of generosity and hospitality, the Buddha of compassion. So the commandments may seem simply to be deontological, but in practice for the most part they are an exercise in virtue ethics.

Detractors are wont to tell us that all religions have different moral codes so 'they can't all be right'. So it has been solemnly pointed out that while men in synagogues are supposed to wear hats, men in Christian churches must go bare-headed – which shows (the argument has it) that both sets of rules are quite arbitrary. But hats or no hats is just a matter of manners. Manners have moral connotations because they show respect or the lack of it – but they are not at the core of morality. More deep-lying are the taboos which, so some economists say, are typically rooted in the economy of particular societies. So the deep reason why Jews and Muslims are forbidden to eat pork is that pigs are woodland creatures – not suitable for desert peoples at all.

But what matters most in all religious ethics is the underlying attitude: and the attitudes that all the great religions demand of us is always the same. *All* preach personal humility, and all teach what the Buddhists call compassion and Jesus called love. I suggest that these two – personal humility and compassion (and particularly compassion) – are indeed the most fundamental notions or feelings that underpin all moral codes, of everyone, whether they deem themselves to be 'religious' or not. We could (and I believe should) add a third: a sense of reverence towards all life and towards the universe as a whole.

These attitudes do not spring simply from a 'rational' view of the universe, but they all make perfect sense in the light of modern knowledge. The modern philosophy of science tells us above all that the universe in the end is beyond our ken and omnipotence is beyond our reach: all that talk of 'conquering' nature is not only vile but ludicrous. Humility in the face of nature is the only sensible course. Compassion is rooted (surely?) in our fundamental, innate sociality, for sociality requires unselfishness, and unselfishness is best reinforced by a true concern for the feelings and welfare of others. The sense of reverence is probably felt most strongly by those who have a sense of transcendence – who feel in their bones that there is more to the universe than meets the eye; and that that 'more' may include an embedded consciousness, and indeed a sense of direction and purpose.

In short we might argue that morality, far from being arbitrary, reflects the way that human beings really are – and the way the universe really is. The universe is fundamentally cooperative, and cooperativeness is made manifest in sociality, underpinned in sentient creatures like us by a positive *desire* to be social; and this desire requires compassion.

There is a theoretical fly in this ointment, of course (there always is). So it was that David Hume also pointed out that 'is' is not 'ought' (or this, at least, is a paraphrase of what he said). This means that although it may indeed be 'natural' to behave compassionately (although this runs counter to the crude neo-Darwinism that has prevailed of late) this does not mean that it is right. To suppose that 'is' means 'ought' is to fall foul of what the Cambridge philosopher G.E. Moore in the early twentieth century called 'the naturalistic fallacy'. So it was that when Humphrey Bogart as Charlie Allnut in *The African Queen* told Katherine Hepburn that it's in a man's nature to get seriously drunk (on warm gin in his case) she replied with sublime peremptoriness that 'Nature, Mr Allnut, is what we are put in this world to rise above'.

But how do we square Hume and Hepburn with the teaching of the Roman Catholic Church, which emphasises 'natural morality'? At least, the Catholics most strongly condemn behaviour that they conceive to be 'unnatural'. Thus the Catholic orthodoxy objects to contraception not primarily because they are killjoys but because, they say, it frustrates God's purpose: for the natural purpose of sexual congress is to beget children. But if it isn't necessarily right to do what nature intends, then why is it wrong to go against its wishes?

Cardinal John Henry Newman in the mid nineteenth century had the answer to this. He conceded that there is no logical, straight line connection between what is natural and what is good. But, he said, in real life (as opposed to the mathematician's drawing board) there may often be no single thread of logic between one idea and another – but the two ideas may be related anyway. In the same kind of way, the mightiest ships are tied to the quayside, perfectly securely, by ropes. Yet there is no single fibre in the ropes that runs all the way from the ship to quay. Each rope is compounded from a myriad different fibres, all individually rather feeble, but all pulling in the same direction and, when taken together, mighty strong. We could say, indeed, that Newman anticipated the twentieth century concept of non-linearity.

By the same token, moral good is not necessarily related by any single thread of reasoning to what is natural. But a thousand different hints and clues proclaim that acts that seem natural are at least more likely to be good than those that seem to fly in the face of nature. Those hints and clues include reasons such as those offered by the utilitarians – for utilitarian thinking isn't all bad: merely inadequate. Mainly, though, the clues are provided by our own intuitions, our twinges of conscience. We tend to feel most uncomfortable with acts that seem most unnatural; and we have no good reason to doubt our intuitions, for they are evolved, honed by the universe itself in the interests of survival; and in the end, the key to survival is harmony.[34] Of course, nature does much that seems bad. In many species, infanticide is routine; and the way that cats 'play' with mice, and orcas with sealions, seems far from salutary. But nature nonetheless is in essence cooperative, and thread by thread we are led from cooperation to compassion. All in all, then, we don't need to 'rise above' nature. We simply need to be in tune with it, and with our own natural selves.

9. Metaphysics, Science, And Religion

Confusion reigns in the western world, indeed in most of the world, and it's not surprising. At least until the last few centuries people's lives everywhere were generally guided by some religion or other, and each religion provided a complete worldview. Each religion had a view on what the world is really like, and how it got here, which are matters of cosmology; and each one told us how we ought to behave, which are matters of morality; and each assumed that it was party to the truth, which in general was relayed to the people at large (including the political rulers) by the priesthood, who were assumed to have special insight. Overall the worldview provided by each religion was and is coherent. In the most widely practised of all modern religions, Christianity, the same God who made the whole universe also laid down the rules of behaviour and so the morality itself has a cosmic dimension and everything we want and need to know forms a coherent whole (even if one or two points are a little tricky).

Different religions have often come into contact, one with another – sometimes jogging along happily enough side by side, at least for some decades or centuries; sometimes locked in bitter conflict; and sometimes achieving some measure of synthesis, as Christianity has managed to do with many a pagan practice worldwide. But although different religions are obviously different in detail, and this has sometimes led to serious conflicts, they all share a common belief in the huge idea of transcendence – the idea that there is more to the universe than meets the eye, and in particular that the universe is driven by an intelligence and has purpose, and that human beings ought to behave according to that purpose. Many throughout history have doubted all this – it seems there have always been sceptics and atheists – but the overall sense of transcendent purpose and cosmic moral order has surely prevailed everywhere.

But the rise of western science since the late Middle Ages and the particular contribution of the eighteenth century Enlightenment, with its emphasis on 'rationalism', has posed a threat not simply to particular religions, but to the whole idea of religion. It has led many to question whether there is any agenda at all in the universe beyond what we can see, touch, and measure, and so it has undermined the whole concept of transcendence and hence of purpose and of cosmic moral order. Not all scientists and not all Enlightenment thinkers were or are out-and-out materialists and atheists, not by any means, but the scientific-rationalist agenda has nonetheless provided a formal framework for materialism and atheism that was not there before. Worse: some defenders of materialist-atheism have become as evangelical and sometimes as bellicose as any religious zealot. They are not content simply to question the premises of religion. They are wont to declare that those premises have been shown to be nonsensical, and that the practice of religion and belief in religion are positively destructive, not to say evil, and that they are holding back both the material and the political and moral progress of the human species. Evangelical atheists, in short, want religion done away with. Stalin and Mao were prominent among political leaders whose anti-religious wrath waxed most mightily; and in modern Britain and doubtless elsewhere a mixed bag of intellectuals have been waging a war of words that is every bit as vehement, and perhaps for some people just as wounding, as the physical attacks of dictators.

All this is a huge pity. It would surely be a terrible thing if everyone thought and behaved the same way (cultural diversity matters for all kinds of reasons, including reasons of long-term survival) but it is a terrible thing, too, that there should be such conflict that is so obviously destructive. Yet humanity surely is missing a huge trick. It is obvious to everyone that the world's religions have many regrettable features of many kinds, but it also seems obvious that in principle they are wonderful and necessary. Science too is wonderful and necessary, and the perceived and largely contrived battles between science and religion that seem to be raging worldwide are ludicrous: obviously destructive but also, for the most part, intellectually feeble in the extreme, even though some of the most conspicuous combatants are among the most acclaimed of modern intellectuals. A worldview is possible and necessary that embraces both science and the core ideas of religion. More than that: a worldview that did embrace them both would not be an *ad hoc* piece of fiction, devised to make us feel better and to keep the peace. A worldview that truly embraced them both would, I suggest, take us as close to truth as it is possible for human beings to approach.

This worldview – this different view of life – would not be just one more religion, to add to the many thousands that are out there already. It would embrace the core principles of all of them. Science would not be presented as it so often is nowadays as the only reliable guide to the truth, and as the rival to all worldviews that do not follow its rules. We would acknowledge the things that science is good at, and recognise the things that it is not good at, and once science was put in its place it would enhance the grand worldview.

The task of devising this grand, all-embracing worldview falls to metaphysics. It is what metaphysics is for. That is why it is such a huge pity, as Seyyed Hossein Nasr has said, that metaphysics has more or less gone missing from western thinking. In so far as metaphysics does survive in the world as a whole, it tends to be parceled out between different religious traditions, and so the way of thinking that ought to be the great integrator, has itself become fragmented. Human beings seem to have a gift for fragmentation. Perhaps it's the Renaissance legacy, or perhaps it's a tribal thing.

So far this book has sketched out some of the key elements of the grand world view that seems possible and desirable. In this chapter I want to start putting the bits together again. We can start by looking at one of the great vexations of our age, indeed of the past few centuries: the uneasy relationship between science and religion.

Science and religion

Prominent among the great modern myths, and very destructive as myths so often are, is that science and religion have always been at loggerheads, and that this is inevitable. A shortlist of alleged historical narratives is wheeled out in support. Thus we are told that at the start of the seventeenth century the Catholic Church led by Pope Urban VIII tried to put a stop to the astronomical speculations of Galileo Galilei – the church presented to us as the stick-in-the mud defender of ancient dogma, and Galileo as the harbinger of a new and better age in which the shackles of superstition would be shaken off. More generally, we are given to understand, the Catholic Church rejects the whole idea of scientific inquiry, the demystification of God's mystery.

From a later century we are given to understand that Darwin's great idea of evolution by natural selection was greeted with rapture by rationalist scientists and condemned by god-bound clerics, with Darwin's polemical friend T.H. Huxley trouncing the dim-witted Bishop 'Soapy' Sam Wilberforce in a showdown in Oxford's newly founded Natural History Museum in June 1860 (a mere seven months after *Origin of Species* was published). A further trouncing followed in July 1925 at the trial of the State of Tennessee *v.* John Thomas Scopes, held in the town of Dayton – the famous 'Monkey Trial'. John Scopes, a local schoolteacher, was prosecuted for teaching evolution. Defence lawyer Clarence Darrow reduced the stick-in-the-mud religious fundamentalist and bigot William Jennings Bryan, chief witness for the prosecution, to a blustering wreck. Or at least, this is the story presented to us in 1955 by Jerome Lawrence and Robert Edwin Lee in their play, *Inherit the Wind*, later filmed with Spencer Tracy as Darrow (though his name was changed to

Drummond) and Fredric March as Bryan (though in the film he was called Brady). Tracy as ever was the archetypal reasonable man, down-to-earth with sleeves rolled up yet wise as Solomon, while March was perhaps the greatest of all screen blusterers. In essence, the Monkey Trial rages on, and a most unedifying spat it is.

All in all, we are often told, the two worldviews of science and religion are simply incompatible. Some seek to make peace but the best they generally achieve is to divide human contemplation between the two of them, with science answering the 'how' questions, and religion addressing the question 'why'. As the evolutionary biologist Stephen Jay Gould put the matter in *The Hedgehog, the Fox, and the Magister's Pox,* published posthumously in 2003, science and religion are two 'non-overlapping magisteria' – where a 'magisterium' is an area of intellectual authority.[35] In other words, at best, science and religion have the relationship of bookends, holding the world between them. At worst, they assail each other from separate promontories, with a vast gulf in between; a gulf that one might imagine as a barren desert, or as a raging torrent, but definitely as no-man's-land.

All of which, I suggest, is the most awful junk.

To begin with, the standard tales from history, at least as they are commonly told, have drifted a long way from the truth. As Arthur Koestler described in *The Sleepwalkers* in 1959, Urban VIII was a serious intellectual, and he and Galileo began as good friends. Urban did not object to scientific exploration, and he certainly did not condemn Galileo out of hand. But there were points of difference between the accepted cosmology of the Catholic Church (which traditionally saw the Earth as the centre of the Solar System), and the heliocentric model proposed by Galileo, and since it was and is the Pope's job to uphold Catholic teaching, Urban felt that there should at least be some attempt to reconcile the two viewpoints. The Church came up with a solution: it required Galileo to acknowledge that his revolutionary notions were simply a hypothesis. The term 'hypothesis' did not have quite the same meaning at the start of the seventeenth century as it does now, but a modern scientist, or at least one versed in the philosophy of science, would surely have leapt at such an offer. *Of course* the idea was a hypothesis. All science is compounded of hypotheses of greater or lesser scope and certainty. The point was simply to ensure that Galileo's notion was taken seriously, as the Pope was clearly prepared to do.

If Galileo had accepted this formula then the perceived science-religion rift might never have come about, or at least become so entrenched; and the whole course of western history would have been different. But Galileo was obstinate, not to say pig-headed, and instead of accepting what seems to have been an eminently acceptable compromise, he wrote a play in which he seemed to be presenting Urban as a simpleton. Popes in Urban's day were both religious leaders and significant statesmen, and all Europe in the early seventeenth century Europe was on permanent amber alert, anticipating trouble of a religious and political kind which indeed broke out at regular intervals, usually in several ways at once. It was obviously foolish to mock so publicly one the most prominent players in European affairs, whose own position was so delicate (what with the breakaway Church of England and the rise and rise of Protestantism and the ever-fractious cardinals). In the end, the trouble that came to Galileo wasn't at all bad: a few years of comfortable house-arrest, looked after by his (deeply devout) daughter.

The general notion that the Vatican is in principle anti-science seems simply to be untrue. In the nineteenth century some of the monasteries of Europe served virtually as scientific research centres; and so it was that Gregor Mendel worked out the basic principles of genetics in the garden of the Augustinian monastery of St Thomas at Brno in Moravia, which is now in the Czech Republic. There he enjoyed a great deal more freedom to pursue his ideas than most modern scientists are allowed. Then he became the abbot. The modern Vatican employs its own scientists, among them the Jesuit Guy Consolmagno, a former president of the International Astronomical Union, who has very engagingly described his own work and the relationship between science and religion in *God's Mechanics*.[36]

The notion that scientists welcomed Darwin's evolutionary ideas is belied by contemporary evidence. Often it's the peripheral accounts that make the point – one of which is Mrs Humphry Ward's novel of 1888, *Robert Elsmere*, in which the eponymous hero renounces his position as a Church of England vicar and becomes an Evangelical teacher. He remains no less devout, but he has abandoned the traditional Christian theology. What's interesting here is that Darwin's ideas on evolution have absolutely no influence on his decision to change horses. He is happy (as many clerics were) to accept that God was perfectly entitled to allow his creatures to evolve by natural selection, and indeed that this was, and is, a marvellous way to ensure that all are well adapted to their conditions and to each other. He abandons the standard Church of England theology of his day (though not his Christian faith!) for reasons that have to do with Christology – the study of the historical Jesus, which in the late nineteenth century was proceeding apace. Mrs Humphry Ward was a serious thinker from a donnish and ecclesiastical family, and knew of what she wrote. She must have been a keen participant in many a discussion both on theology and on Darwin.

The particular tale of Huxley routing Wilberforce at Oxford turns out to be seriously reconstructed. Those present at the debate report that there was so much noise that no-one could hear what was being said, and Wilberforce felt afterwards that he had won the debate. Wilberforce reviewed *Origin of Species* and Darwin was impressed by his understanding and felt that his criticisms were very fair, and needed answering. As philosopher Michael Ruse has described, not least in *The Evolution Wars* of 2001, the Scopes Monkey Trial in general, and Bryan/ Brady in particular, have been seriously misrepresented, while *Inherit the Wind* can reasonably be seen as a work of libel. For example, in the play (and the film), the lawyer Darrow/Drummond forces Bryan/Brady to admit, with gasps all round, that the seven days of the Creation may not literally have been days, each of twenty-four hours, but could have been of indefinite length. In reality, Bryan offered this point himself, entirely unprompted. In reality Bryan was a serious thinker, a pacifist and reformer who opposed the corporates and big banks, and stood as a Democratic Presidential Candidate.

As for Darwin's own religious views – well, no-one can be sure. But he was buried in Westminster Abbey on April 26, 1882, just a week after his death, and this is what Dean (Frederic) Farrar, said by way of eulogy:

> This man, on whom for years bigotry and ignorance poured out their scorn, has been called a materialist. I do not see in all his writings one trace of materialism. I read in every line the healthy, noble, well-balanced wonder of a spirit profoundly reverent, kindled into deepest admiration for the works of God.

Indeed, Darwin intended at one point to take holy orders, and so would have joined the honourable succession of naturalist-clerics who include John Ray and the Reverend Gilbert White (whose *Natural History and Antiquities* of 1789 was one of Darwin's favourite books) – and many more. At various times throughout his life Darwin said that he had lost his faith – but not because of his ideas on evolution. He could not reconcile the suffering of his own father in later life, and then the death of his daughter Annie in 1841, with the idea of an all-merciful God. But also, I feel, Darwin simply found it hard as many Victorians did (including the semi-fictional Robert Elsmere) to accept traditional nineteenth century Church of England theology. Certainly, throughout *Origin of Species,* the 'spirit profoundly reverent' shines through. Some say he wrote in this vein to please his wife and placate his many devout friends but there seems to be no worthwhile evidence at all for this. I think he wrote as he felt.

In truth, as John Hedley Brooke relates in his excellent and indeed seminal *Science and Religion,* the relationship between science and religion these past few centuries has not been uniformly hostile, not by any means. Sometimes it has been positively synergistic – with theologians often posing the problems that the scientists felt they should address. Indeed the central question that Darwin addressed – how is it that each creature on this Earth is so well adapted to its own particular circumstance? – was originally proposed as a matter of theology. The traditional answer was that we should thank God's providence. However, while Genesis suggests that each creature must have been pre-planned and put in place in an already perfected form, Darwin suggested a more subtle and much more efficient process: essentially, installing prototypes, and allowing them to interact and to adapt to the physical conditions and to each other, and to go on adapting; the whole global ecosystem gradually unfolding and becoming more intricate. Pre-planning would not work. Evolution by means of natural selection obviously does.

Most tellingly, the 'scientific revolution' of the seventeenth century was religiously inspired. As Professor Peter Harrison of Oxford University relates in *The Fall of Man and the Foundations of Science*, traditional belief had it that when God created Adam, he gave him knowledge of all things (except the forbidden knowledge of good and evil); but this was lost when Adam and Eve were expelled from the Garden of Eden. A significant succession of thinkers beginning in the Middle Ages and extending through to modern times then argued that the task for humankind was to recover this pristine knowledge. As Francis Bacon put the matter at the start of the seventeenth century, the purpose both of science and of philosophy was to see how human knowledge –

> ... might by any means be restored to its perfect and original condition, or if that may not be, yet reduced to a better condition than that in which it is now.[39]

So it was that a key text from Genesis, the book so often mocked by post-Darwinian materialist-atheists, to a significant extent inspired the entire scientific agenda of the modern West.

All the acknowledged founders of modern science from the seventeenth century – Galileo, Newton, Descartes, Leibniz,

Robert Boyle and John Ray and all the rest – were devout. They felt that to use their God-given intellects to explore God's own reasoning was itself an act of worship. They felt about their science just as J.S. Bach, a few decades later, said of his music: that it was all for the glory of God. All of them took God's existence more or less as a given – although Descartes, like some medieval theologians before him, was also at pains to demonstrate that God's existence is *necessary*: that the universe cannot be explained otherwise. When Newton spoke of natural 'laws' he meant God's laws, for he took it to be self-evident that there could be no laws without a law-maker. Only in later times did anyone conceive that the laws of Nature might have arisen, as it were, spontaneously.

I once chaired a meeting on religion and science where a woman declared that to expose God's mysteries in the way that scientists try to do is blasphemous – but to me such comment itself seems blasphemous (if we are to bandy such words at all). To suggest that God's mysteries *can* be exposed and hence made trivial is to imply that God is just some kind of conjuror, whose tricks really do look tawdry once you know how they are done. But the universe is not like that. The more that scientists have looked, the deeper the mysteries become. Socrates said that the more philosophers know, the more they realise they don't know; and the same is true of science. In truth, the primary role of science is not to achieve omniscience, as is sometimes so crassly implied, or to provide high technologies that would lead us to omnipotence. It is to enhance appreciation; and the more we appreciate, the deeper the mystery becomes.

Science has at least intriguing things to say about Darwin's own dilemma – one that has troubled religious thinkers particularly in the Christian tradition for two thousand years. If God is really all-merciful, and is also all-powerful, why do so many bad things happen in the world? If God really wants bad things to happen, then he cannot be all good. If he does not want bad things to happen and they happen anyway, then he cannot be all-powerful. The particular strand of theology that addresses this problem is known as theodicy. In truth, the problem of theodicy (why bad things can happen in a world created by a benevolent all-powerful God) is mainly a Christian concern. Other religions don't necessarily start with the concept of a single all-powerful, all-beneficent God – and certainly not with one who is answerable for his actions to human beings – so the issue does not arise.

Yet there are plausible answers to the Christian dilemma nonetheless. Some of the bad things that happen on this Earth can clearly be blamed on human beings – murder, genocide, and all the rest. All these result in the end from the exercise of human free will. Yet if we did not have free will, but were merely God's puppets, then we would not be the great creation that we are. Human miscreancy, paradoxically, is the price that must be paid for the possibility of achieving human greatness. The task for humanity, and the ultimate perfection, is to *choose* to behave well.

Natural disasters such as volcanoes, earthquakes, and tsunamis require a more subtle explanation – one that can be supplied through the insights of twentieth century science. Before that, natural philosophers tended to believe as Laplace did in a clockwork universe in which everything is in principle predictable – including all future earthquakes; raising the question of why God made things that way. But now we know that uncertainty is built in to the fabric of the universe. At the smallest scale it is impossible to predict which way a photon will jump. On the larger scale we know for example that the crust of the Earth is in constant motion – the continents are carried this way and that on shifting tectonic plates. Sometimes the plates become locked and then jerk free and then there is a quake and perhaps a tsunami – but there is no knowing when this can happen.

So why make the world in such an eccentric fashion? One explanation is that the movement of the tectonic plates is vital for continued life on Earth. It recirculates the vital minerals on the grandest scale. Without tectonic stirring, the Earth would soon be sterile as the moon. More generally, as Oxford physicist Paul Ewart has pointed out, an entirely pre-determined clockwork universe would be incapable of innovation. It would simply grind on (so all computer models suggest) until it ground inevitably to a stop. It's only because of the built-in uncertainty that change and continuation are possible – such that the universe as a whole can evolve harmoniously. The bad things that result from all this restlessness and uncertainty are the price we pay for a universe that actually functions, and can support life, and go on supporting life. On the grand scale of things, it's a small price. As Professor Ewart puts the matter, 'Chance is necessary'. No one can claim that these arguments are actually true, and still less that they represent the whole truth and nothing but the truth. But they are certainly plausible and they certainly make sense, and it is just possible that if Darwin had had access to twentieth century notions of uncertainty and non-linearity and had not felt quite so hemmed in by the rigours of nineteenth century theology, then he might never have said that he had lost his faith, and his name could not have been co-opted to the cause of atheism.

Finally, just to round off this part of the discussion, I love the following quote from Samuel Taylor Coleridge:

> I have heard it said than an undevout astronomer is mad. In the strict sense of the word, every being capable of understanding must be mad, who remains, as it were, fixed in the ground in which he treads – who, gifted with the divine faculties of indefinite hope and fear, born with them, yet settles his faith upon that, in which neither hope nor fear has any proper field for display. Much more truly, however, might it be said that, an undevout poet is mad: in the strict sense of the word, an undevout poet is an impossibility ...[40]

Yet we cannot properly describe the relationship between science and religion without discussing what they actually *are* (and what they are not). I have discussed what science is in earlier chapters. So what is religion?

What is religion?

All religions have an intellectual agenda – which is basically that of metaphysics. That is, virtually all religions seek to supply a complete and coherent account of all that is, which means they must address the three fundamental questions: 'What is the universe really like?' (with the sting in the tail, 'How come?'); 'How do we know what's true?'; and 'What is good?'

But religions aren't merely an exercise in metaphysics. All of them embed the metaphysical abstractions within a narrative that in part is generally geared to the history and experience of particular groups of people – tribes, nations; and in part (and more importantly) relies upon the teaching of one or a number of key authorities, including prophets and gurus, sometimes modified later by priests and theologians. Finally (and, many would say, most importantly) religions are concerned with *practice*: how we should live our everyday lives, and the observation of special ceremonies and rituals.

In detail, different religions can look very different, and those who seek to disparage the whole idea of religion like to emphasise the differences. After all, the detractors say, since they all say different things they can't all be true; and since they recommend different ways of life then all their recommendations must be arbitrary. But, as always, what really matters is the essence – and the essence, I suggest, is the same in all of them. All produce very similar answers to the three key questions of metaphysics. At least, they may differ in cosmology – Buddhism, for example, envisages time to be cyclic, while in the Abrahamic tradition it is more linear (or at least, the thread of time that human beings live in seems linear). But in matters of morality the greatest of the acknowledged teachers – Moses, Jesus, Mohammed, Buddha, Confucius – made essentially similar recommendations, and lived their lives accordingly. Religious rituals differ but all share the same fundamentals of private contemplation (which may or may include prayer) and of collective worship.

So, in more detail…

The cosmology of religions

Most religions that are commonly recognised as such invoke the idea of a God, or of gods – supernatural beings who control things; although some religion, notably the three great Eastern traditions of Buddhism, Daoism and Confucianism, are 'non-theistic', without a specific God or gods. But *all* religions invoke the idea of transcendence. They take it to be the case, or indeed to be self-evident, that there are forces at work in the universe, an overriding intelligence, beyond and above what we can see directly, and touch and measure. God, or gods, may be seen as a personification, or a reification, of this fundamental idea.

But out-and-out materialists reject the idea of transcendence. For them, what we see is what there is. The logical positivists tried to justify this by suggesting that any idea of anything that cannot be seen, touched, or measured is nonsensical, since ideas that are not simply those of logic, or amenable to scientific experiment, are mere whimsy (or as A.J. Ayer said, 'gibberish'). But logical positivism does not work. It is clear, too, that explanations of a purely materialist kind do not adequately explain the facts of the universe (including the fact of consciousness).

None of this demonstrates, of course, that there really is a transcendent dimension to the universe. But it strongly suggests that such an idea is perfectly plausible, and perfectly reasonable, and very possibly necessary. You may feel in your bones that this is not so, as some materialist-atheists presumably do. But it is a mistake to suppose that the idea of transcendence has been shown to be wrong, or ever can be shown to be wrong, and there is no good reason (and I stress *reason*) to reject the idea of transcendence out of hand. The question for the undecided is not whether you are persuaded by Richard Dawkins' arguments, but whether you are prepared to rank Dawkins's bone feelings above your own. If so – why?

On an intriguing point of detail: Richard Dawkins has said that if the God of Abraham (of the Jews, Christians, and Muslims) really was the Creator, then he leaves much to be desired. The world's myriad creatures are well adapted overall but they nonetheless have serious shortcomings. So it is that the human immune system, vital to ward off pathogens and parasites, plays so many nasty tricks, including allergies and auto-immune diseases of many kinds, possibly including multiple sclerosis and various forms of arthritis. Humans walk upright well enough – but have terrible problems with their backs and knees (although they might not have done in their hunting-gathering days, when they exercised more and did not live so long). Before the days of modern obstetrics, human childbirth was positively dangerous – and this is not just an artifact of civilisation, for women in hunter-gatherer tribes may also suffer mightily. In short, says Professor Dawkins, God was a lousy engineer.

But I like the riposte from the Catholic theologian Tina Beattie (in *The New Atheists: The Twilight of Reason and the War on Religion)*. For, asks Professor Beattie, why assume that creation means engineering? Artists create too, do they not? – and artists do not, as engineers are assumed to do, simply produce a design and then build the thing that they have designed. They have an iterative relationship with their creation. They have a general idea of where they want to go but once the artist sets to work then they and their artistic creation operate in dialogue. First the artist sketches his or her idea and then sees what the sketch suggests should happen next; then modifies the work in the light of where the work itself seems to be leading; and so on and so on. In principle there is no end-point, except when the artist chooses to call a halt. We can see this process unfolding in the unfinished works of all artists – painters, musicians, novelists; the work becoming richer with each re-working as new ideas present themselves – yet always staying roughly within the original conception (until some fresh idea arrives out of the blue and suggests a complete change of tack). This model of artistic creation seems perfectly to reflect what we see in the evolutionary record; and perfectly explains the imperfections that we may choose to perceive.

Incidentally, too, engineers in practice do *not* simply turn their idea into nuts and bolts and present it to the world as a *fait accompli*. Their first attempts are called prototypes, in urgent need of refinement; and when the creation is finally marketed it is ruthlessly subjected to the rigours and demands of the market, and modified and modified again, exactly as Darwin perceived the workings of natural selection. So it was that the bicycle was first invented in principle more than two hundred years ago but has been modified more radically and in more ways in the past twenty years than at any time before. Yet I thought that the bike that I rode at age fifteen was the last word (and so perhaps did the makers). In art, in engineering, in evolution, there is no last word. There is only perpetual striving.

The Welsh journalist and broadcaster John Humphrys once commented that the most fundamental question, which distinguishes the religiously inclined from the non-religious, is 'Do you believe in God?' But this is not the most fundamental question. The idea of transcendence is more basic than any particular idea of God. Belief is a tricky concept, too. Personally I know that my head is full of all kinds of ideas, but I am hard pressed to say precisely which of them I *believe*. I only know that I find some ideas very attractive while others I feel are simply irritating, if not downright daft.

So the most fundamental and informative question, I suggest, is not: 'Do you believe in God?' but 'Do you take seriously the idea of transcendence?' – or, even more generally, 'Are you prepared to take seriously the idea of transcendence?' I think all members of all recognised religions must answer 'Yes' to this, and so would a great many other people of a kind who would claim to be 'spiritually' inclined even though they belong to no particular religion. Between the two, that certainly accounts for most of the human race. Yet for various reasons human beings collectively behave as if we do not take transcendence seriously – as if we were all out-and-out materialists, like the economists who first dreamed up the rules by which we now live, and the politicians who now enforce them. In behaving as we do, we are not being true to ourselves. We have been persuaded or coerced into behaving against our own natural inclinations. If only we saw that we have been led astray, and were brave enough to follow our own intuitions, the world really would be a much better place.

How do we know what's true?

Detractors, including atheists and general gainsayers, are wont to tell us that religion is 'irrational'; that it depends on 'blind faith'; or, as the chortling Lewis Wolpert has put the matter, that it requires us to believe six impossible things before breakfast. Wolpert is an emeritus professor of biology at University College, London and a semi-professional atheist, who writes, lectures, and broadcasts a great deal on religion. But he never progresses beyond parody. Science is easily parodied too – but parody won't do. The stakes are far too high.

To be sure, religion relies very heavily on intuition. So does science, as we have seen: but religion is much more upfront about this. Indeed we might say that the principal *method* by which religion arrives at its special insights is by the careful cultivation of intuition, and ultimately upon experience: an overwhelming realisation that such-and-such an idea, or such-and-such a way of looking at the world, is the way things really are. In this context I love the word 'prehend', meaning grasp. Bertrand Russell and his friend A.N. Whitehead used the term in a philosophical context to mean the grasping of ideas – although for some reason 'prehend' is missing from the Oxford English Dictionary, or at least from my edition of it (although 'prehension' is in there). Intuition leads us to prehend ideas; to grasp them all of a piece. I think that what religious people do, and what prophets certainly do, is to put themselves in a state of receptiveness in which the mind is freed from everyday concerns and clutter, and grasps what is out there. If we take on board the idea of universal consciousness then we can say more grandly that the receptive mind seeks to engage directly with that universal consciousness. This I am sure is what mystics do. This is the reason why prophets and gurus throughout the ages have sought quiet places – mountain tops and deserts and the shade of trees. The most creative scientists do the same (such as Barbara McClintock, and all those astronomers through their long lonely nights). So of course do artists. The idea is to allow intuition to happen.

This process is indeed 'non-rational' – if rationality is defined in the manner of logicians and logical-positivists, as a buttoned down mathematical analysis of empirical evidence. But it certainly is not 'irrational', with its implication of bonkersness. Cultivated intuition leading to prehension is a vital component of all creative thinking, and without creative thinking we are forever stuck with whatever is known already. There is an analogy here perhaps with Paul Ewart's suggestion that all serious advance in the universe as a whole depends on the random and unpredictable leap (or with the evolutionary idea that change depends on random genetic mutation). It's clear whatever way we spin it that if we truly seek to know what is true, then we *must* cultivate intuition. That, indeed, becomes the only 'rational' thing to do. It may seem paradoxical that it is rational to cultivate non-rational thinking but that is the case nonetheless. All creative people know this, whether in theology or the arts or in science. But some scientists and philosophers have chosen to pretend that is not so, and mock the very process that makes all creativeness possible. There are certain areas, too, where mere 'rationality' – empirical measurement and logic – cannot supply the answer. One of them is the key issue of whether or not there is a transcendent dimension to the universe. In such matters, where measurement and logic fail, cultivated intuition must be the principal guide.

Yet anyone who knows any theology knows that theologians are not content merely with intuitive insight. Ideas based on intuition are subjected to the most rigorous *logical* analysis. Theologians, not least in the form of Jesuits, have included some of the sharpest wranglers that humanity has ever produced. The parodists are wont to tell us that medieval theologians were content simply to fantasise – but, as so often, the opposite is the case. From the tenth century onwards, thanks to Arab translators and thinkers, western theologians became increasingly aware in particular of the work of Aristotle – and became increasingly obsessed by logic. Thus St Anselm in the eleventh century sought to demonstrate on *logical* grounds that God must exist. He argued that since God was the best that could be conceived and the best that could be conceived must exist – for anything that did not exist could not be the best. The argument doesn't sound good to modern ears and some took issue with it even his own time – but the refutation is a lot harder than it looks. Thomas Aquinas in the thirteenth century tried to show that God must exist by appealing to Aristotle's idea of the unmoved mover, the prime cause of everything – again an essentially logical approach.

Even if such arguments are flawed (which most theologians and philosophers now feel is the case) they are certainly not irrational, or even non-rational. But they do serve to demonstrate yet again that for some questions even the most buttoned-down rationality may fall short. Overall, the history of theology emphatically is not one of unrestrained intuitive fantasising. It is and has always been a dialogue between cultivated, refined, non-rational prehension, and the sharpest possible analysis. Science at its best is the same. In fact, this is the way that human beings think, and other animals too, presumably, and very efficient it is too.

Detractors say too that religion depends on 'blind' faith. Some religious people have also unfortunately spoken of the 'leap of faith', giving atheistic polymath Dr Jonathan Miller the opportunity to ask, also in chortling vein, what kind of 'athletic training' is needed to make such a leap.

But 'faith', like most big words of this kind, does not mean what a lot of people take it to mean. It certainly does not require us to believe impossible things for the sake of it, as Professor Wolpert supposes. It does imply that we are prepared to take seriously propositions that have not been nailed down beyond all possible doubt, and possibly never can be nailed down (for nothing can, in the end). When faith is defined thus, we see that we are all faith-driven – no-one more so than scientists. The belief that science is all we need to investigate the world, and indeed that it will lead us to omniscience, is an absolute statement of faith. So are the many points of detail within that broad assumption: the idea, for example, that we really can 'lick cancer' (as US President Richard Nixon put the matter in the 1970s). So indeed as many a philosopher including David Hume has pointed out, is the broad supposition that the laws of physics will be the same tomorrow as they are today. But faith in any context need not be 'blind'. There can be good reason to have faith: intuition supported by a million observations, some and probably most of which are made unconsciously. Thinking Christians and Muslims and Jews, and Sikhs and Jains and Buddhists, give a host of very good reasons why they have faith. Even if the detractors remain unconvinced by those reasons, it is simply untrue to suggest that their faith is 'blind'.

Neither is it true that faith must be achieved by some single athletic 'leap', in the manner of St Paul on the road to Damascus. Many significant thinkers came to religion slowly. The Christian writer and don C.S. Lewis claimed to be an atheist at age nineteen, and became a Christian step by imperceptible step over the following ten years. St Augustine went all around the intellectual houses before winding up as one of the principal figures in the history of Christianity, very significant both in Roman Catholicism and in the main branches of Protestantism. As often as not, as Francis Bacon put the matter:

> It is true, that a little philosophy inclineth man's mind to atheism; but depth in philosophy bringeth men's minds about to religion.[41]

Finally, and with a touch of desperation, hard-nosed detractors are wont to tell us that there is 'no evidence' for any transcendent intelligence behind the surface attributes of the universe. Professor Wolpert likes to say there isn't a 'shred' of evidence. But people who say such things don't know what 'evidence' is, any more than they know what faith is, or for that matter what rationality is. For evidence is not a signed affidavit, a known and irrefutable fact. If it was, there would be no need for courts of law. Judges could be replaced by pocket calculators. 'Evidence' is merely a fact or an observation that is deemed to be compatible with a particular proposition.

Is it really the case that there are no facts compatible with the idea of a transcendent intelligence? In truth there are so many facts suggesting this, that most people through most of history have taken this to be self-evident. As Dan Dennett and Richard Dawkins have both been at pains to point out, the purely materialist interpretation of Darwinian evolution that has become standard is seriously counter-intuitive: it is hard to believe that so many various creatures, each so exquisitely designed, could have come about without a designer. People believe in the existence of a designer because the facts seem to suggest this; and that, by definition, is evidence. The problem is to provide evidence that this is not the case. This, of course, is impossible. We can demonstrate only, to some people's satisfaction, that the variety of intricate creatures *can* be explained in purely materialist terms, if we put our minds to it. But to say that life's existence and variety can be explained in purely materialistic terms in no way demonstrates that such an explanation is actually true, and certainly not that it represents the whole truth and nothing but the truth. The materialists argue that materialism is the default position: that the burden of evidence lies with those who want to show otherwise. But there is no good reason at all to accept this. We could just as well argue that the burden of evidence lies with the materialists, who are defying widespread human intuition.

What is good?

Detractors are wont to tell us that different religions have different moral rules – so those rules must be arbitrary and again, they can't all be right. So Richard Dawkins has pointed out that while men must remove their hats in Christian churches (apart from the senior clergy) Jews must keep their heads covered. And so on.

But such quibbles miss the point. Hats or no hats, shaking hands or bowing, these are merely manners: in effect, local conventions. Manners are socially important and they do have a moral content (they show respect, which implies concern for the feelings of others) but they are not what is meant by 'morality'. Deeper than manners lie various taboos: don't eat pork; don't work on the Sabbath. These have more gravitas than manners and excite more opprobrium when flouted – but still they are in the realms of local tradition (often with an economic basis). What should be meant by 'morality' are the deep rules that should shape all human behaviour.

Here again we tend to be misled. For we are commonly given to understand (not least by the people who espouse particular religions) that the moral rules within each religion are simply those laid down by God, or at least are as laid down by some transcendentally inspired profit or guru – and hence that the moral rules of religion are essentially deontological: a duty to obey orders. But this again, surely, misses the point. For in principle the moral rules of religions – *all* religions – are primarily those of virtue ethics. We are not told simply to obey particular rules, but to develop particular *attitudes*: to each other, to life in general, to God (or indeed to the universe as a whole). Those attitudes are as described earlier: personal humility; compassion (which the Christians tend to call Love); and a sense of reverence – towards God, or towards the Universe as a whole.

The very considerable bonus, as discussed throughout all the earlier chapters, is that the fundamental virtues as perceived by all religions are perfectly in accord with modern views on animal psychology and sociology, and hence in line with modern science. They are rooted in cooperation, and life and the universe are essentially co-operative. At least the basic ideas of virtue ethics are in line with *truly* modern science: not with the eighteenth century Enlightenment/logical positivist/ materialist/ ultra-Darwinian version of science that is conventionally conceived to be modern.

Of course we should be personally humble, and indeed humble as a species, when we acknowledge, as the modern philosophy of science tells us, that life and the universe must forever be beyond our ken, and beyond our control – except here and there, ad hoc, in the short term.

Naturally compassion is not only good, but eminently sensible. Life is fundamentally cooperative, and sociality works, and sociality in thinking creatures depends on empathy, generosity, and indeed altruism – a sense that other people and the society as a whole really matter. It is morally right to care about others (which implies some measure of self-sacrifice) but it is also in our own self-interest to create an atmosphere of amity. Biology tells us too that all other species are our literal relatives; and that many of them are conscious and aware – and on both counts they are truly our fellow creatures.

The sense of reverence springs from a simple appreciation of the wonder of it all (a sense that science, properly conceived, makes stronger and stronger); and the feeling of wonder becomes even greater, I suggest, when we take seriously the sense of transcendence – the notion that there is even more to it all that science is able to tell us. To be fair, many scientists including some who are dyed-in-the-wool materialists also emphasise the sheer wonder of nature. This is the theme of Richard Dawkins's latest book *The Greatest Show on Earth* (Transworld, 2009). Dawkins, though, does not generally speak of reverence but of awe. Reverence implies love and gratitude, and a desire to be part of. Awe, intentionally or not, has overtones of fear, as in the 'shock and awe' of the US military. Reverence is the thing. Reverence seems to me to imply a sense of transcendence.

Most of the scientists that I have met became scientists because they were and are moved by this sense of overwhelming wonder, and the reverence or awe to which it gives rise. It is certainly what lured me into biology, more than half a century ago. Many were drawn into science by compassion – the belief that science when properly applied can improve human life and help to conserve life as a whole. Many scientists (including of course the long and continuing tradition of scientist-clerics) embrace the idea of transcendence. Others feel transcendence in their bones but perhaps because of their hard-nosed education and for fear of their hard-nosed colleagues suppress that feeling. Some of the latter claim merely to be 'agnostics'. But in some, in the manner that Sigmund Freud discussed in a slightly different context, the thing that is denied, the suppressed intuition, bubbles over, to find expression in atheism, sometimes of the most vehement kind. Or so it seems.

The end of conflict

In truth, most of many spats between science and religion are no such thing. Most are battles between materialism and atheism on the one hand, and particular theologies on the other. But as we have seen, there is no reason – and I stress *reason* – why a scientist should be a dyed-in-the-wool materialist atheist, and many of the greatest scientists have been and are deeply devout. Contrariwise, the grand concept of 'religion' should not be equated with particular theologies.

In fact you can be very religiously inclined, or 'spiritual' if you prefer the term, without embracing the details of any particular theology. The essential components of religion as I see things are fourfold: a sense of transcendence, a sense of oneness, and a morality based on compassion and humility. All are underpinned by rationality and intuition in dialogue: 'cultivated intuition'. If you take the theology out of any of the great religions you are left with these four principles, with greater or lesser degrees of emphasis (Christianity, for example, despite St Francis and a long tradition of naturalist-clerics, does not seem to emphasise oneness). But what we have left when the theology is gone is still worthwhile. Indeed what is left is very similar to the religions and philosophies of the East: Buddhism, Taoism, and Confucianism, and indeed Shintoism. Given that there is such tension between all the religions of the world (and most of the traditional ones are simply ignored by all except their immediate followers), why don't we simply cultivate a metaphysic that embraces the essence of them all? I would like to see such a metaphysic at the centre of all education, and the subject of public gatherings and ceremonies. This metaphysic would not be intended to replace the established religions – far from it. But it could be seen as the foothills that could lead whoever is inclined to the peaks of the great religions; but if people preferred to stay in the foothills – well: that would be enough.

Metaphysics, though, taken on its own, isn't all that the world needs. It is 'necessary but not sufficient'. Truly to make a difference in the world we need to devise systems of politics and of economics, and day-to-day ways of living, that are in tune with the metaphysical ideas – transcendence, oneness, compassion, humility. Again, the notion that metaphysical musing must be translated into way of life is a central precept of all the Eastern religions. How we might achieve this in the modern world is the subject of the next and last chapter.

10. Renaissance

The world needs changing, and the change must begin with the *Zeitgeist,* with the ideas that fill the basement of our minds. Do we believe, as the neo-Darwinian scientists and the neoliberal economists have been assuring us these past few decades, that life is and must be an all-out battle from conception to the grave? Do we agree with the premise that lies behind the prevailing economic theory, that our happiness will increase as we grow richer and more powerful? Do we agree with Gordon Gekko that 'greed is good'? Do we *want* the world to be like this? Are we content with it?

Or do we feel intuitively that these, the ideas that now dominate our lives, are seriously misguided, and damaging? So why don't we trust our intuitions, and act on them? For in the end, as David Hume observed in *A Treatise of Human Nature:*

> Reason is, and ought to be only the slave of the passions, and can never pretend to any other office than to serve and obey them.[42]

Our 'passions' – intuitions – are telling us (aren't they?) that what we have now won't do. It's striking too that Hume, one of the greatest of all rationalists, tells us that our passions *ought* to lead; but then, like all the greatest rationalists (Einstein, Socrates, Thomas Aquinas, Darwin, Newman are among those that come to mind – and of course the 'Romantics' like Samuel Taylor Coleridge) he recognised the limitations of mere rationality. Our intuition tells us that the moral underpinning of our modern economy, and hence of our lives, is junk; and we should be prepared to trust our intuitions, even though intellectuals are wont to tell us not to. But in truth, too, as we have seen throughout this book, a growing body of scientific literature is telling us (as a few scientists have always insisted) that the 'life-is-one-long-punch-up' thesis is wrong. Life is fundamentally cooperative. That is not just wishful thinking. That is the way things are. Out of a sense of cooperativeness and the deep desire for conviviality comes compassion. We cannot reliably be cooperative unless we truly give a damn about other people, and other creatures. Cooperation and conviviality are the names of the game. It is a huge serendipity that the best long-term survival tactic, both for individuals and for the human species as a whole, is to be *nice*.

We have been told too these past few decades that science is marching on at such a pace that soon we'll understand everything; and out of this omniscience will come omnipotence – we'll be able to fashion the world as we want (and we are given to understand that what we *want* is more and more material goods). So to neo-Darwinism and neoliberalism we are invited to add technophilia, for only high tech can solve our problems, or so we're told. It's just a question of paying for it. So we are advised above all to get rich, so that we can buy the technologies that will dig us out of the holes that we have dug ourselves into by attempting to do so.

But all this too is serious junk. High tech is wonderful (I would not be without my word processor, or the internet, and am personally very grateful to the modern pharmaceutical industry) but, like science itself, high tech does only what it can do. It will always leave us far short of omnipotence partly because the forces at work on us can be too great but also because the science that lies behind high technologies must always leave us far short of omniscience. As Socrates said, the more we know the more we ought to realise how little we know. In the end, life and the universe will always be mysterious. This realisation – a re-discovery of what sages and people at large have known ever since human beings first started thinking – ought to breed humility. It is *rational*, in the light of uncertainty, to proceed cautiously. Anything else is bravado.

We may feel too, intuitively, that there is more to the universe than meets the eye, or can ever meet the eye; and this, I suggest, is what is meant by a sense of transcendence, and is often called 'spirituality'. We may or may not choose to equate this sense of transcendence with the idea of a God or gods. We may or may not feel (as physicists including Peter Russell suggest) that the transcendent dimension of the universe can be equated with a universal intelligence. But however we may choose to develop the idea, many a survey (and many a conversation) tell us that most human beings do have a sense of transcendence. Many people (including me) take this to be obvious, and so they have throughout history.

Finally, and in some ways perhaps most importantly, there is the notion of life's 'one-ness'. This sense of one-ness lies at the core of most religions, and is supported absolutely by modern biology; by the physicists' observation that we all share the same atoms in one never-ending exchange; by Darwin's idea that everything that lives on Earth shares the same ancestor; by the ancient and modern idea that the intelligence we share with other creatures is part of the fabric of the universe. This notion is the antidote to anthropocentricity. Those who believe that all life is one cannot reasonably believe that human beings are the only species that counts, and that we have the right to treat the rest as a resource. But that is what the present *Zeitgeist* assumes. In the prevailing jargon, life is called 'biodiversity' and valued only for what it can do for us: 'ecotourism'; 'eco-services'; 'the environment' (meaning real estate). The vocabulary itself is vile.

Yet throughout the world's greatest literature, and in people's heads and hearts, we already have the elements of a *Zeitgeist* that really could serve us well: a morality based on compassion and humility; a cosmology that preferably would include a sense of transcendence; and a feeling for life's one-ness. I suggest that *most people* would find this agreeable, and that many if not most would accept it as a perfectly good description of how the world literally is. This worldview rests heavily on intuition – but then: anyone who does not take intuition very seriously is all too likely to be an *idiot savant,* and such people can be very dangerous. In any case, sound biology tells us that our intuition is rooted in our evolved response to the problems posed by the universe this past 3.8 billion years, and it is not obvious why it should lead us astray. Our own rationality tells us that our intuition should not be lightly overridden. Rather it should be cultivated – by contemplation and in dialogue with rationality, in the manner of theologians and artists.

Such a worldview should change our personal attitude to other people and to all of life, of course; and this is vital. Yet it is not enough, for as Karl Marx put the matter (the words are inscribed on his grave at Highgate cemetery in London):

> Philosophers have interpreted the world in various ways,
> the point however is to change it.

So we must translate the new (but ancient-rooted) metaphysics into action – and in practice this is happening, all over the world. Personal desires to create a different world are being transformed into grass-roots movements of all kinds. They have many different agenda – new forms of energy, better food, better schools, the restoration of landscapes and crafts and so on – but through them all (or at least the ones that really matter) we can trace a common thread. They all aim to give all of us, ordinary Joes, control over our own lives – to restore what ought to be our birthright. Commonly these grass-roots movements begin with groups of friends in the pub. Sometimes, as recommended by Italy's Slow Food Movement, these social gatherings develop into more formal 'convivia', which then form links with others of similar mind. Some of these groupings grow into formally recognised non-government organisations, or NGOs. In *Blessed Unrest,* Paul Hawken calculates that there may now be a million different NGOs worldwide fighting all manner of causes. Although they are so various in make-up and intent, through most of them runs a common theme: that of human progress through cooperation, in the cause of conviviality; the precise opposite of the ultra-competitive winner-takes-all mentality that now prevails. Between them these many and various initiatives form what is loosely called the 'civil society'.

To change the world – not just superficially, for a year or so, but to bring about a true sea-change, a 'paradigm shift' – we need to start with a critical mass of people who want to do things differently. But a critical mass does not have to be a majority. Indeed it very rarely is. Twenty per cent of people pulling in the same direction, or even far fewer, can start the sea-change happening. The great serendipity is that we already have the necessary mass – people who want a world that's rooted in cooperation and conviviality. Just ask people what they would prefer, and what they really feel about the world and how it should be; or, indeed, tot up the number of recognised NGOs. But numbers alone are not what counts. A mass does not become critical unless it coheres. The task, then, is to achieve coherence. So how?

Coherence

Coherence must run in two directions – both vertically and horizontally.

Vertical coherence, I suggest, may be conceived in three tiers. In the bottom tier, the basement of our minds, lies metaphysics: all the grand ideas that contribute to our personal worldview and collectively to the *Zeitgeist*. In the middle tier is what (with a slight shift of metaphor) might be called the engine room of a formal society: economic and political theory. At the top are the day-to-day realities of life; the trades and crafts and professions we must pursue to keep ourselves alive and in good spirits: farming, manufacture, health care, building, teaching, academic theorising, story-telling, and all the rest.

Horizontal coherence means that the different ideas and pursuits within each tier should be pulling together.

So how does all this work in practice?

Most of this book has been concerned with elements of the bottom layer, the metaphysical underpinning – but metaphysics has more or less disappeared from formal curricula or from day-to-day thinking except within the context of formal religions where it is invariably coloured by particular theologies. This neglect, says Professor Nasr, is the root cause of all the world's ills, and the more I have thought about this, the more I am sure he is right. In this we should reverse the trend of the past few centuries absolutely. Metaphysics should again be at the heart of formal education and a *leitmotiv* of everyday discourse.

Then the ideas that should form the new metaphysics and at present merely clutter up the basement of our minds need to be tidied up. At present, it seems as if life's most fundamental ideas are at odds with each other. So the neo-Darwinians have told us that life is inveterately nasty, based on self-interest and competition – and this seems to be horribly at odds with our sense of morality: except of course that morality has been re-defined in the manner of Gordon Gekko, so that self-interest is now deemed to be good. So we either have to put up with a contradiction – neo-Darwinian biology is at odds with our basic morality; or we must achieve coherence by twisting morality itself into something foul. But if we acknowledge (as modern biology suggests) that cooperation must trump self-interested competitiveness, then the biology and the morality are perfectly in accord. That is coherence.

Or then again we have been told these past few hundred years and particularly over the past few decades that science and religion are inveterately at odds. But when we construe them properly, see what each of them is and is not, and what each can do and cannot, then we see that in truth they can and should be in perfect harmony. At least, the differences are on points of detail which for the most part can be seen as the difference between accounts that content themselves with what is visible and aspire to be literal, and those that seek the largest possible truths that must be expressed through metaphor. Science should be embedded in a larger metaphysical context, and religions attempt to express that context in terms of particular narratives. Seen thus, religion *embraces* science, when both are properly construed, and in truth they need each other, and we need both. Einstein was above all wise and to quote him once more:

> Science without religion is lame; religion without science is blind.

Overall, then, the essential metaphysics that is found at the heart of all religions, and which could be developed without the trappings of any particular theology, can be summarised in four words: transcendence, oneness, compassion, and humility. The whole is underpinned by intuition cultivated by dialogue with what is commonly called rationality. This is in line with how most people feel, and it reflects (so modern science and its philosophy are telling us) how the world really is. It is perfectly coherent and it surely would serve us well.

Politics and economics should spring naturally from the underlying metaphysics: they are the mechanisms by which we may translate our shared beliefs on what is and what should be, into a formal modus operandi. In contrast (yet again) present-day political and economic theory is free-floating: rooted not in a robust metaphysics but on the arbitrary deliberations of whoever at the time happens to be taken most seriously. For the past few decades the world has been guided by the neoliberal thoughts of Milton Friedman, as first delivered to the world by Mrs Thatcher and Ronald Reagan. So now we have an absurd mismatch between the rhetoric of politicians and the economy that they support. The rhetoric calls for peace, social justice, and what is now called 'sustainability'; but the prevailing neoliberal economy requires us to be at each other's throats and to produce more and more stuff as if the Earth was infinite and there is no tomorrow. This is incoherence writ large. To be sure, in the wake of the 2008 financial crash, even the US government has accepted that it can be necessary to adjust the free market from time to time, and has invested public money in new industries, as recommended by the greatest of twentieth century economists, John Maynard Keynes. But Keynes did not seem seriously to consider the limitations of the Earth itself any more than the neoliberals do. He assumed as the ruling moderns do that high tech can solve our problems – and of course in his day (he died in 1946) there was no intimation of global warming (although the underlying physics was known). So again we need to start afresh from first principles.

Politics and economics from first principles

The prevailing economy, worldwide, as it stands, is a nightmare. The craziness behind it is enough on its own to explain the terrible mess that the world is in. But again, it's not just the mechanisms of the economy that needs to be re-thought. As with science, as with religion, we need to re-think what economics actually *is*.

Economics as it now stands has become just an exercise in money: how to make as much of it as possible. Money itself has become divorced from reality. If we want more money we print it (or rather, since printing is cumbersome, we deem it to exist, and register the amount thus conjured into being in a computer). Currently there is deemed to be about eighteen times more money in the world than there are material goods to buy. So money is no longer a symbol of real things. It is a pure abstraction. The whole mirage is based on debt – the debt economy. Our own wealth is assessed not in terms of what we actually have, which in turn might reflect what we have actually done that is of use, but on how much we can borrow, and in recent years it wasn't even necessary to be creditworthy.

So it is that most of us most of the time now spend most of what we earn on paying back the debts on loans – loans which were more or less forced on us ten or twenty years ago by banks that thought that lending money and charging interest was simply the easy way to get rich. Combine this with the neoliberal market economy in which the *only* imperative is to make more money than anyone else in the shortest time in the interests of competition, and in which morality itself is decided by what people are prepared to buy, which in turn depends on how much they can borrow; and in which the prevailing battle-cry from the most powerful governments is for more 'growth', meaning more money year on year, forever and ever; and we have a recipe for disaster that might have been concocted by Mephistopheles himself.

I won't go on. Now that the nonsense has collapsed, as it was bound to do, everyone has become aware of what before was concealed or we preferred not to know: that apparently the affluence at least of some people over the past few decades has been nothing more than a bubble; a fantasy; a giant exercise in creative accountancy.

But the truly alarming thing is that the same people (or the same kind of people) who presided over this disaster are still in charge. The same nonsensical theories and ambitions still prevail – the debt economy, the neoliberal market, the all-out imperative for 'growth'. The powers-that-be have not learnt the lesson that is so obvious to people at large. They have nothing in their heads except to bring about 'economic recovery', to 're-build' the economy to what it was, and to restore 'confidence' (in money and in the bankers and governments who control it), to get us back to the merry-go-round of debt and ever-increasing consumption. Worse still is that intellectuals of all kinds have been dragged into this. In particular, science itself, once perceived as the noble and incorruptible seeker after truth, has been bought out. Science has become a commodity, paid for by corporate wealth, dedicated to ends that make its sponsors richer; and the scientists who are best-paid and most secure, and are most likely to appear on television and tell the rest of us what's what, are the ones who toe the economic line. Many don't, but most of them are out of work and those that are employed are not in the front line.

The errors begin with the fundamentals. Page 1 of all standard economic textbooks invariably tells us that the goal of economics is to make people happy and that happiness is proportional to wealth and that therefore the role of economics must be to maximise wealth. The nonsense is compounded by the failure to acknowledge: (a) that the world itself is finite; and (b) that our fellow creatures matter. But then, the whole house of cards is underpinned by technophilia – the embedded belief that high tech can always get us out of trouble. We need only to pay for the necessary research, it seems; which brings us back to the notion that our task in life, guided by economists, is to be as rich as possible, to repair the damage that the urge for wealth has brought about.

But in reality the economy is not just a matter of money. At least, economists qua economists should be primarily concerned with money, but they should not be in charge of society. We should not allow their miasmic abstractions to dominate our lives. Instead, all economies should be rooted in reality – what is it necessary to do, and what is it possible to do; and by agreed morality – what is it right to do. These in truth are the fundamental questions of metaphysics. Get the metaphysics right, in short, and the economy should follow. If we got our priorities right, then as Keynes put the matter:

> ... the economic problem will take the back seat where it belongs ... and the arena of heart and head will be occupied where it belongs, or reoccupied by our real problems, the problems of life and human relations, of creation, and of behaviour and religion.

Yet the status quo has bound itself into yet another nonsense. The word has got round, spread by the minority who are doing well out of the present economy that the 'free' in 'free market' and 'free enterprise' implies that all of us are free. The very poor are assured that they could be very rich if only they played their cards right, and many clearly believe this, and vote for the very rich while they themselves may live in trailer parks. For the alternative to the free market, we are told, are the 'controlled economies' that once were seen in Stalin's USSR and Mao's China, and as seen today in beleaguered North Korea. Even the word 'socialist' – the general notion that we all deserve a reasonable crack of the whip – is now taboo. It is not heard these days even at conferences of Britain's 'Labour' party. 'Socialist' is now taken to mean Stalinist, apparently.

But the alternative to the 'free' market is not Stalinist central control. The realistic alternative is what we saw in Britain and the US through most of recent history – what is commonly called 'social democracy'. Free market economics, the forerunner of today's neoliberalism, has sometimes prevailed over the past two hundred years but, for the most part, successive British and US governments have taken it for granted that their principal task was to control the economy in the interests of the society as a whole. That is, they relied upon free enterprise to power the economy, supported by the standard mechanisms of capitalism – judicious loans and interest and all the rest – but they took it to be obvious that this basically capitalist structure should operate within a moral and legal code that worked in the public interest: code that it was the job of government to oversee. So it was for example that the Founding Fathers of the United States in the early nineteenth century, John Adams, Thomas Jefferson, James Madison *et al*, passed laws to restrict the powers of corporates – ruling for example that they had to demonstrate that they were operating in the public interests. The Founding Fathers did not call themselves social democrats, but they demonstrated its essence.

Keynes too pointed out that free markets do not in practice work as well as their advocates claim they do. They are innately unpredictable (chaos theory applies); they are always liable to collapse – and theory as well as experience shows that they do not necessarily recover on their own; and they do not lead automatically to social justice. Self-evidently, it is the government's job to moderate the market in the public interest. If they do not do that, it is hard to see what they are for. A market modified in the interests of social justice is social democracy; and social democracy that also took account of the physical realities of the Earth and recognised the existence of our fellow creatures (in effect, a green social democracy) would at least be fit for purpose.

In practice, social democracy is rooted in the mixed economy. Social democrats do not claim that only private ownership is acceptable, as the modern neoliberals do, and seek to 'privatise' everything in sight as Britain's Tory-led coalition is trying to do. Neither do they attempt, as central-control Stalinists did, to suppress private enterprise in favour of all-out public or state ownership. One of Britain's greatest post-war politicians was Aneurin (Nye) Bevan, was a committed socialist and nowadays would be seen as a dangerous leftie. Yet he in the end was simply a social democrat – as the following makes clear, from his short, personal manifesto of 1952, *In Place of Fear*:

> A mixed economy is what most people of the West prefer. The victory of Socialism need not be universal to be decisive ... It is neither prudent, nor does it accord with our conception of the future, that all forms of private property should live under perpetual threat. In almost all types of human society different forms of property have lived side by side ... But it is a requisite of social stability that one type of property ownership should dominate. In the society of the future it should be public property.[44]

But then, in Bevan's day, the Tories were social democrats too. Harold Macmillan, often seen as the archetypal Tory Prime Minister, spent a great deal on public housing, as did his pre-war predecessor Neville Chamberlain. The difference between Macmillan and Bevan was simply one of emphasis: the ratio of private to public ownership. The Tories did not shift wholesale from social democracy to the neoliberal free market until Thatcher got into her stride in the 1980s. Indeed there was far more difference between Bevan and his communist contemporary Nikita Kruschev, or between Macmillan and Thatcher, than there was between Bevan and Macmillan.

So there is nothing frightening about social democracy and the mixed economy – for why should there be? It is rooted in common sense: the kind of economy that seems to emerge naturally when the leaders seek to reconcile social justice on the one hand, with reasonable efficiency on the other. Common sense relies heavily on intuition: the bone feeling that something is right, or is not. The extremes – the controlled economies of Stalin and Mao, and the neoliberalism we have now – are both abstractions, dreamed up by intellectuals of the kind who abandon intuition in favour of what they perceive to be rationality; the danger that David Hume warned us against. We must beware of intellectuals. Intellectuals continue to tell us that the market economy works (or will do one day) and that high tech can solve all our problems (or will do one day) even though it is already abundantly clear that the game is up. As George Orwell constantly reminded us, intellectuals are prone to cling to ideas long after they have been shown to be foolish.

I think we should all stop taking fools seriously and rely once more on what our instincts are telling us; that we need an economy rooted in biological reality and morality, and that does not defy common sense.

Overall, too, we clearly must take 'democracy' much more seriously than we do. We claim to be 'democratic'. Young men and nowadays young women too are sent to the far corners of the world to fight for 'democracy' – and yet our lives are in some ways more constrained than ever, by micro-management from above. Democracy implies that the rules and structure of society as a whole reflect what people really want and if one of the main arguments of this book is right – that most people are nice most of the time – then a society that was truly democratic ought to be convivial. The term 'democracy' has been interpreted in many different ways but it ought to mean (should it not?) perhaps above all that we, people at large, Ordinary Joes, have control over our own lives. So how?

Taking control

The metaphysics is the foundation. Economic and political theory provide the broad guidelines within which we live. But what matters to most of us most of the time is our day-to-day lives; what we actually do.

Day to day life in practice breaks down in to a series of jobs – the things we need to do to keep ourselves alive: building, health-care, farming, cooking – and so on and so on. All of these *métiers* in societies like ours tend to have rules of their own. In the Middle Ages crafts and trades were overseen by gilds (which had a great deal going for them!). Nowadays each is monitored by its own particular rules and conventions, commonly reinforced by professional bodies and sometimes reinforced by the Trade Unions.

Over all, though, sits the government – and the restraints imposed by the all-pervasive competitive economy. So the final say rests with the powers-that-be, which means with people who are powerful and/or rich – people who have dedicated their lives to power and wealth. Those who make the final decisions are not necessarily expert. The principal goal in all government and commerce these days is to maximise wealth, in line with demands of the maximally competitive global market – so all of our *métiers,* even teaching, health care, and wildlife conservation, are required to pay their way. Of course we have to be aware of budgets. But as Keynes pointed out, big countries can always afford the things they *really* want – so it's mainly a question of wanting the right things. In Britain right now we cannot apparently afford social services but we could apparently afford the Iraq and the Afghan wars and we did afford the Olympic games (which indeed were a tribute to public enterprise and a model of conviviality).

The combination of top-down, often cack-handed control from above and the overweening compulsion to maximise wealth even when this is so obviously inappropriate, has compromised all work. When I was preparing to leave school in the late 1950s, in Tory but social democratic Britain, we took it to be self-evident that the primary purpose of work was, first, to help to create a better world; and secondly, to pursue a way of life that was personally satisfying. Since I was studying biology most of my contemporaries were aiming to be medics (which most of them have become) and this was their guide: public service and personal fulfilment, in harmony. There has never been full employment in my lifetime and perhaps for various technical reasons that is not possible – but it was taken for granted nonetheless that most people would find jobs in one or other of a hundred different professions, crafts, and trades. Many alas simply were obliged to do whatever job came to hand but many, in all pursuits, sought consciously to reconcile public service with personal satisfaction. Life was in many ways fraught and much harder than it now (at least for some people) but overall it was *OK*.

Life these days is in many ways far more comfortable, at least if you're not in the bottom 20 per cent, but it is not OK. Everyone is pressured not simply to do their job well but to conform to all manner of arbitrary rules and to make as much money as possible – or else be taken over by somebody who will. Politicians wring their hands over unemployment and poverty – and yet promote an economy that *demands* that labour should be cut, which means that people should be put out of work, because it is deemed to be more 'efficient' to reduce the workforce, and therefore more profitable, and profit is all. Clearly in the short term companies can save money by putting people out of work but in the longer term they tend to fail through under-manning and of course the overall cost of unemployment and the consequent waste of human potential is horrendous – and that's only in cash terms. If we added in the human misery and sheer waste of life that doesn't get costed, as we strip out the labour force to improve 'the bottom line', we find that we are striking a very poor bargain indeed.

So we need to get away from the ruling mentality – to replace the simplistic belief in accountancy and high tech, with compassion and common sense. But the people in charge, the powers-that-be, the drivers of governments, corporates, banks, and their attendant intellectuals and experts, are committed to the status quo. It is what has made them powerful and rich. So if we want a better world then we, people at large, Ordinary Joes, have to take matters into our own hands. We have to create true democracy. So how?

In principle there are three ways to bring about change. The first is by Reform. The rules are tweaked and tweaked again until the desired endpoint is reached, more or less painlessly. In practice this means going cap and hand to people in power and asking them nicely to change their ways. This can work up to a point. It has to be tried. But it is not enough. This is shown very clearly by reference to the area that I have been most directly involved in for the past forty years and which I feel in the end is the most important: food and farming. The great supermarkets that now in effect control the world's food supply grow rich by scouring the world for the greatest bargains, playing farmers worldwide against each other, then imposing the greatest mark-up they can get away with. To achieve economies in a world where oil is still relatively cheap they demand that farming should be practised on the largest possible scale – monocultural fields of wheat as far as the eye can see; factory farms with thirty thousand cows or a million pigs or poultry (literally) – so that the 'farms' can provide mega-quantities of completely uniform produce to be borne in jumbo trucks or sometimes in jumbo planes to central distribution points then on to the points of sale. Factory farms need very few workers so millions of farmers are thrown out of work, and finish up in urban slums (almost one third of all city dwellers). Oil is burnt prodigiously. Forest is felled to make room. Yet the point of the whole exercise is not to produce good food for everyone (and in truth in the modern system a billion are going hungry) but to maximise profits for the shareholders, and produce a caste of super-rich executives.

In truth to feed the world well we need a network of mixed farms that are known to be the most productive per unit area, and are by far the most sustainable. Such farms are complex and so must be skills intensive – lots of farmers as opposed to the fewest possible; so there is no advantage in scale-up; so they should in general be small to medium sized. Such farms are best suited to local distribution, as provided by traditional shops and stalls (and nowadays in the west increasingly by farm shops and farmers' markets, though these put too much strain on the farmer).

There is no plausible route by which the mega, trans-national corporates that own the big supermarkets can be transformed, step by step, into the kind of farming and distribution system that the world actually *needs*. The factory farms and the supermarkets are the wrong model. They cannot become the kind of institution the world needs without committing hara-kiri. But governments love corporate control because it is easy to deal with; because it seems to maximise profits and so increases GDP and so allows governments to claim 'economic growth', which they now see as their *raison d'être;* and because (let's face it) the big transnationals that modern politicians love to encourage commonly offer them places on the board. Indeed, ministers of agriculture in Britain in recent years have sometimes been chosen precisely because they have big industry connections. This is presumed to give them expertise.

So although reform can achieve a certain amount, it cannot do what really needs doing. It cannot transform the present power structure, the corporate-government complex, into the kind of institutions that promotes justice and conviviality. What else is there?

The second route to change is Revolution. But all-out revolution is too messy. There is too much collateral damage, and revolutions very rarely produce the results that their instigators intend. All political action is non-linear in its effects but in revolution, non-linearity applies in spades. Besides, the world now is so precariously placed that the mess that would result from all-out revolution could prove terminal. Well-directed activism is often necessary, however: and surprisingly often, the law proves to be on the side of the activists. Much of what the powers-that-be are now preventing us from doing is, in fact, perfectly legal (if you appeal to the right bits of the law).

The third route to change is Renaissance – and this is the one that can really work: the one that can produce the turn-around that's needed, radical and permanent, without creating mayhem. The Italian Renaissance of the fifteenth century onwards has a lot to answer for as discussed in earlier chapters but it does provide an excellent model for cross-the-board change.

Renaissance literally means re-birth. It implies that all we have to do to change the world is to start doing things differently *despite* the powers-that-be. All over the world, in all areas of life, Renaissance is happening. In all fields – alternative energy, food, housing, education – groups of people are taking their own affairs into their own hands, beginning as outlined above with chats between friends that become convivia that may grow into NGOs. There is a whole catalogue of mechanisms, essentially borrowed from capitalism, that can help to turn these initiatives into publicly recognised endeavours with real clout – all legal, but all helping to create a culture that does not take its lead from governments which in turn are beholden to the corporates and banks. Chief among these mechanisms are community interest companies (CiCs) and Industrial Provident Societies (IPSs), and cooperatives of many kinds. Worthwhile initiatives can be supported by 'ethical investment': people investing, just as they might in the conventional companies of the stock exchange, but more concerned to do something worthwhile than simply to achieve the greatest return. All this, collectively forming 'the civil society', are exercises in cooperation; and all are motivated by values that are not those simply of material wealth.

These principles can be applied, and are being applied, in all branches of our existence – building, transport, education, what you will. The literature is vast and growing and I cannot summarise all that is happening here; but Martin Large's recent *Common Wealth* provides an excellent lead in. Again, though, my own special interest of food and agriculture illustrates the principle of Renaissance particularly well. At present, I and associates are promoting what we are calling 'Enlightened Agriculture', sometimes shortened to 'Real Farming': farming that is 'expressly designed to provide everyone with good food, forever, without wrecking the rest of the world'. This ambition is perfectly realisable, despite what we are told by the powers-that-be. As suggested above, the basis of enlightened agriculture is *not* the industrial monocultural megafactory, but the low-input maximally mixed, skills-intensive farm that in general would be small to medium-sized. We have established a website, www.campaignforrealfarming.org, where people who share these ideals (and especially farmers) can swap ideas and find ways to work together (and the website has already produced some useful liaisons). Every year as part of the campaign we run the Oxford Real Farming Conference, where good ideas are shared communally.

Recently, together with an ethical bank and several brokers who specialise in versions of ethical investment and other forms of fund-raising, we have set up the Fund for Enlightened Agriculture. People at large are invited to invest (or where appropriate to donate) to support the kind of farms and smallholdings that are appropriate, and also to support related enterprises – including the small shops or people's supermarkets that small mixed farms are suited to. Eventually we hope to establish what I am currently calling the Trust for Enlightened Agriculture. Like the present-day National Trust, the farming trust would raise money from people at large to acquire land – but would dedicate that land in perpetuity to enlightened agriculture: to farming designed to provide good food, with justice, without cruelty, and wildlife-friendly. Eventually, too – quite soon, in fact, with luck – we hope to establish a College for Enlightened Agriculture where all the necessary ideas may be developed and passed on and where we can help to create a new generation of farmers. Since farming affects everything else that we do, and is affected by everything else that we do, the College curriculum must be broad. It must include the feeding of pigs and the raising of wheat but also the state of the world and nature of the economy – and the underlying metaphysics. It would be good, too, to establish a School of Metaphysics – in the hope that in the fullness of time metaphysics will be reinstated at the heart of all education.

The necessary metaphysics (just to say it one last time) is encapsulated in the four key concepts of transcendence, oneness, compassion, and humility. All are at the heart of all religions worthy of the name. All find support in truly modern science and the philosophy of science. Between them they define the attitude that could lead us to create a far better world than we have now – convivial and secure; able in principle to keep our descendants and our fellow creatures in good heart for the next million years, or in effect forever. The clock is ticking, but Renaissance is still possible.

Endnotes

1. Richard Dawkins, *River Out Of Eden: A Darwinian View Of Life,* Phoenix, London, 1996, p.155.
2. Peter Atkins. From 'The limitless power of science' in *Nature's Imagination – the Frontiers of Scientific Vision,* 1995, p.125. Quoted by John Lennox in *God's Undertaker,* Lion, Oxford, 2007, p.8.
3. Thomas Robert Malthus. *An Essay on the Principle of Population,* Chapter VII, p.61, Oxford World Classics reprint.
4. Quoted by Ruth Padel in her introduction to the 2009 Vintage edition of Darwin, p.xiv.
5. A.A. Michelson. *Light Waves and Their Uses* (1903), pp.23–4.
6. Charles Darwin, *Origin of Species,* p.587. Quoted in *The Portable Darwin,* edited by Duncan M. Porter and Peter W. Graham, Penguin, London, 1993, p.149.
7. Darwin, Charles (1845), *Journal of researches into the natural history and geology of the countries visited during the voyage of H.M.S. Beagle round the world, under the Command of Capt. Fitz Roy, R.N.* (Second ed.), London: John Murray, pp.403–4.
8. From Robert M Young: 'The impact of Darwin on conventional thought' in *The Victorian Crisis of Faith,* SPCK, London, 1970, p.26.
9. Gregor Mendel, *Versuche über Pflanzen-Hybriden* (Experiments on Plant Hybridisation) 1865: Published in Verh. Naturforsch. Ver. Brünn *Proceedings of the Natural History Society of Brünn* 1866, volume 4, pp.3–47. First published in English in 1901 in the *Journal of the Royal Horticultural Society* (vol.26, pp.1–32).
10. From T.H. Huxley's *Evolution and Ethics, Prolegomena;* quoted by Matt Ridley in *The Origins of Virtue,* Viking, London, 1996.
11. Matt Ridley, *The Origins of Virtue,* Penguin Press Science, London, 1997, p.24.
12. Richard Dawkins, *The Selfish Gene,* Oxford University Press, Oxford, p.2.
13. Matt Ridley, *The Origins of Virtue,* Viking, London, 1996.
14. Quoted by Matt Ridley in *The Origins of Virtue,* p.18. From Paradis, J. & Williams, G.C., *Evolution and Ethics: T.H. Huxley's Evolution and Ethics*

with New Essays on its Victorian and Sociobiological Context, Princeton University Press, Princeton, 1989.
15. *Nature,* February 23, 2012, vol. 482, p.461.
16. Richard Dawkins, *The Selfish Gene*, Oxford University Press, New York, 1976, p.21.
17. Denis Noble, *The Music of Life*. Oxford University Press, 2006, p.12.
18. In truth, we need not suppose that the ability to recognise other individuals *necessarily* implies intelligence, A recent report in *Nature* (vol. 481, January 12, 2012, p.154) tells us that paper wasps of the species *Polistes fuscatus* recognise each other's faces. But there is independent evidence to suggest that in mammals and birds at least the ability to recognise faces does correlate with general intelligence.
19. Konrad Lorenz. *King Solomon's Ring*. English edition: Methuen, London, 1961. From the introduction.
20. Konrad Lorenz, *King Solomon's Ring* Translated by Marjorie Kerr Wilson, Methuen, London, 1961, p 147.
21. Frans de Waal, *The Age of Empathy,* Souvenir Press, London, 2009, p.12.
22. David G. Rand *et al. Nature,* September 20, 2012, vol. 489, p.427.
23. David Kahneman, *Thinking, Fast and Slow,* Allen Lane, 2011.
24. Anne Frank, from her diary: July 15, 1944. In *The Diary of Anne Frank, The Critical Edition*, Netherlands State Institute for War Documentation, Doubleday, New York, 1989. Quoted by Thom Hartmann, *Unequal Protection,* Rodale Press, 2004, p.1.
25. Geoffrey Miller, *Spent,* Heinemann, London, 2009.
26. George Orwell, 'Notes on Nationalism'. Essay published in *Polemic* in 1945.
27. Richard Wilkinson & Kate Pickett, *The Spirit Level,* Allen Lane, 2009.
28. Peter Atkins. From 'The limitless power of science' in *Nature's Imagination – the Frontiers of Scientific Vision*, 1995, p.125. Quoted by John Lennox in *God's Undertaker,* Lion, Oxford, 2007, p.8.
29. Peter Russell, *From Science to God,* New World Library, California, 2002.
30. Amit Goswami, *The Self-Aware Universe,* Tarcher/ Putnam, New York, 1995, p.10.
31. Amit Goswami, *The Self-Aware Universe,* Jeremy P. Tarcher/Putnam, New York, 1995 p.10.
32. Richard Dawkins, *River Out Of Eden: A Darwinian View Of Life,* Phoenix, London, 1996, p.155.
33. Quoted by Peter Coates in *Ibn 'Arabi and Modern Thought*, 2002, p.61.
34. We might argue, though, on a point of detail, that the Catholics are wrong about contraception – because their biology is bad. For it ought to be obvious that in a species like ours, as in many primates, the biologi-

cal function of sex is at least as much social – bonding the pairs and the groups – as it is procreative. This is most conspicuous in bonobos, previously known as 'pygmy chimpanzees'. But that is a diversion.

35. Stephen Jay Gould, *The Hedgehog, the Fox, and the Magister's Pox,* Harmony Books, New York, 2003.
36. Guy Consolmagno, *God's Mechanics,* Jossey-Bass, San Francisco, 2008.
37. Quoted in 'The Impact of Darwin on Conventional Thought' by Robert M. Young in *The Victorian Crisis of Faith*, edited by Anthony Symondson, SPCK, London, 1970, p.26.
38. John Hedley Brooke, *Science and Religion,* Cambridge University Press, Cambridge, 1991.
39. Peter Harrison, *The Fall of Man and the Foundations of Science,* Cambridge University Press, Cambridge, 2007, p.1.
40. From Lecture VIII, 1811–12. Quoted in *The Portable Coleridge,* ed. I.A. Richards, Viking Penguin, 1978, p 397.
41. From 'On Atheism', in *Francis Bacon, The Major Works*, Oxford University Press, Oxford, 1996, p.371.
42. David Hume, *A Treatise of Human Nature*, 1739. ed. L.A. Selby-Bigge, 2nd edition, rev. P.H. Nidditch, Clarendon Press, Oxford, 1978, p.415.
43. Quoted by Archie Mackenzie in *Faith in Diplomacy*, Caux Books, 2002, p.200.
44. Aneurin Bevan, *In Place of Fear.* Quoted by David Marquand in *Britain Since 1918,* Phoenix, London, 2008.

Further Reading

This is not an exhaustive bibliography – just a personal list of books and authors who for me have been particularly influential.

On evolution and ecology

Stapp, Jan *The New Foundations of Evolution,* Oxford University Press, Oxford, 2009.

Conway Morris, Simon *The Crucible of Creation: the Burgess Shale and the Rise of Animals*, Oxford University Press, Oxford, 1998.

Lovelock, James *Gaia: a New Look at Life on Earth*, Oxford University Press, Oxford, reprinted 2009.

de Waal, Frans *The Age of Empathy*, Souvenir Press, London, 2010.

Darwin, Charles *The Origin of Species by Means of Natural Selection,* ed. J.W. Burrow, Penguin Classics, London, 1982.

—, *Journal of researches into the natural history and geology of the countries visited during the voyage of H.M.S. Beagle round the world, under the Command of Capt. Fitz Roy, R.N.* (Second ed.), London: John Murray, 1845.

Cronin, Helena *The Ant and the Peacock*, Cambridge University Press, Cambridge, 1993.

Noble, Denis *The Music of Life*, Oxford University Press, Oxford, 2006.

On the philosophy and history of science and the nature of reality

Goswami, Amit *The Self-Aware Universe: how consciousness creates the material world,* Jeremy P. Tarcher/ Putnam, New York, 1995.

Harding, Stephan *Animate Earth*, Green Books, Dartington, 2006.

Russell, Peter *From Science to God: a physicist's journey into the mystery of consciousness,* New World Library, California, 2002.

Medawar, Peter *Pluto's Republic*, Oxford University Press, Oxford, 1958 (reprinted 1984).

Ruse, Michael *The Evolution-Creation Struggle*, Harvard University Press, Cambridge, Mass., 2005.
Sheldrake, Rupert *The Science Delusion,* Coronet, London, 2012.
Clarke, C.J.S. *Reality Through the Looking-Glass: Science and awareness in the postmodern world,* Floris, Edinburgh, 1996.
Whitehead, Alfred North *Science and the Modern World*, Cambridge University Press, Cambridge, 1926.
Koestler, Arthur *The Sleepwalkers: A History of Man's Changing Vision of the Universe,* Hutchinson, London, 1959.

On metaphysics, and the relationship of science and religion

Brooke, John Hedley *Science and Religion*, Cambridge University Press, Cambridge, 1991.
Luscombe, David *Medieval Thought*, Oxford University Press, Oxford, 1997.
Coates, Peter *Ibn 'Arabi and Modern Thought*, Anqa Publishing, Oxford, 2002.
Nasr, Seyyed Hossein *Man and Nature, The Spiritual Crisis in Modern Man,* Kazi Publications, Chicago, 1997.
Symondson, Anthony (ed) *The Victorian Crisis of Faith*, SPCK, London, 1970.
Lennox, John C. *God's Undertaker: has science buried God?* Lion, Oxford, 2007.
Wilson A.N. *God's Funeral*, John Murray, London, 1999.
Tarnas, Richard *Passion of the Western Mind*, Pimlico, London, 1991.
Kumar, Satish *You Are, Therefore I Am,* Green Books, Dartington, 2002.
Hutton, Richard *The Triumph of the Moon*, Oxford University Press, Oxford, 1999.
Beattie, Tina *The New Atheists: The Twilight of Reason and the War on Religion*, Longman and Todd, London, 2007.
Harrison, Peter *The Fall of Man and the Foundations of Science*, Cambridge University Press, Cambridge, 2007.

On politics and economics

Illich, Ivan *Tools for Conviviality,* Caldar & Boyars, London, 1973.
Large, Martin *Common Wealth, For a Free, Equal, Mutual, and Sustainable Society,* Hawthorn Press, Stroud, 2010.
Hawken, Paul *Blessed Unrest*. Viking, London, 2007.
Marquand, David *Britain Since 1918*. Phoenix, London, 2008.

The following books of my own are also pertinent

Tudge, Colin *Good Food for Everyone Forever,* Pari Publishing, Pari, 2011.
—, *The Secret Life of Birds. Who they are and what they do,* Allen Lane, London, 2008.
—, *The Secret Life of Trees,* Allen Lane, London, 2005, Penguin Books, London, 2006.
—, *The Variety of Life: A Survey and a Celebration of All the Creatures That Have Ever Lived,* Oxford University Press, Oxford, 2000.
—, *The Day Before Yesterday,* Jonathan Cape, London, 1995, Published in the US as *The Time Before History*, Scribner, New York, 1996.

Index

Related books

The Intuitive Way of Knowing

A Tribute to Brian Goodwin

Edited by David Lambert and Chris Chetland

Professor Brian Goodwin was a visionary biologist, mathematician and philosopher. Understanding organisms as dynamic wholes, he worked to develop an alternate view to extreme Darwinism based solely on genetic factors. His evolutionary vision was centred more on archetypes than ancestors and cooperation rather than competition. He died in July 2009.

This book honours both his work and personal memories of a much loved and respected colleague. Contributors include: Stuart Kauffman, Lewis Wolpert, Fritjof Capra, Margaret Boden, Michael Ruse, Fred Cummings, Mae-wan Ho, Philip Franses, Stephan Harding, Nick Monk, Claudio Stern and Johannes Jaeger

Grow Small, Think Beautiful

Ideas for a Sustainable World from Schumacher College

Edited by Stephan Harding

This book is a collection of essays from Schumacher College on sustainable solutions to the current global crisis. Themes include the importance of education, science, Transition thinking, economics, energy sources, business and design, in the context of philosophy, spirituality and mythology.

The contributors include Satish Kumar, Jules Cashford, Fritjof Capra, Rupert Sheldrake, James Lovelock, Peter Reason, Gideon Kossoff, Craig Holdrege, Helena Norberg-Hodge, Colin Tudge, Nigel Topping and many others.